Lecture Notes in Biomathematics

Managing Editor: S. Levin

7

Michael C. Mackey

Ion Transport through Biological Membranes

An Integrated Theoretical Approach

Springer-Verlag
Berlin · Heidelberg · New York 1975

Author

Dr. Michael C. Mackey
McGill University
Department of Physiology
P.O.Box 6070
Station A
Montreal, Quebec
Canada H3C 3G1

Library of Congress Cataloging in Publication Data

Mackey, Michael Charles, 1942-
Ion transport through biological membranes.

(Lecture notes in Bio mathematics ; 7)
Bibliography: p.
Includes index.
1. Biological transport--Mathematical models.
2. Membranes (Biology)--Mathematical models.
I. Title. II. Series. [DNLM: 1. Biological transport. 2. Ions--Metabolism. 3. Cell membrane permeability. W1 LE334 v. 7 / QH601 M157i]
QH509.M3 574.8'75 75-35923

AMS Subject Classifications (1970): 92A05

ISBN 3-540-07532-1 Springer-Verlag Berlin Heidelberg New York
ISBN 0-387-07532-1 Springer-Verlag New York Heidelberg Berlin

Printed in Germany
Offsetdruck: Julius Beltz, Hemsbach/Bergstr.

TO NANCY

PREFACE

This book illustrates some of the ways physics and mathematics have been, and are being, used to elucidate the underlying mechanisms of passive ion movement through biological membranes in general, and the membranes of excitable cells in particular. I have made no effort to be comprehensive in my introduction of biological material and the reader interested in a brief account of single cell electrophysiology from a physically-oriented biologists viewpoint will find the chapters by Woodbury (1965) an excellent introduction.

Part I is introductory in nature, exploring the basic electrical properties of inexcitable and excitable cell plasma membranes. Cable theory is utilized to illustrate the function of the non-decrementing action potential as a signaling mechanism for the long range transmission of information in the nervous system, and to gain some insight into the gross behaviour of neurons. The detailed analysis of Hodgkin and Huxley on the squid giant axon membrane ionic conductance properties is reviewed briefly, and some facets of membrane behaviour that have been revealed since the appearance of their work are discussed.

Part II examines the foundations of electrodiffusion theory, and the use of that theory in trying to develop quantitative explanations of the observed membrane properties of excitable cells, in particular the squid giant axon. In addition, an *ad hoc* formulation of electrodiffusion theory including active transport is presented to illustrate the qualitative nature of cellular homeostasis with respect to intracellular ionic concentrations and membrane potential, and cellular responses to prolonged stimulation.

Part III consists entirely of personal efforts, and collaborative work with M.L. McNeel, to examine the molecular foundations of electrodiffusion theory and arrive at a (hopefully) more realistic treatment of transmembrane ion movement. The work described in this last part is not complete, and I hope that others will be encouraged to extend and/or modify the analysis in a number of obvious ways.

The material of Parts I and II forms the basis of a final year undergraduate course that I have given for the past two years in the Department of Physics at McGill, and which was offered in the Department of Physiology before that.

I would like to thank American Elsevier Publishing Company for their permission to reproduce much of the material in Mathematical Biosciences (1975, 25, 67-80) in Chapter 9; Robin Barnett and Eugene Benjamin for computations on which some of the figures of Chapter 10 are based; and the Rockefeller University Press for permission to reproduce material from the Biophysical Journal in Chapters 11 through 15. My special thanks go to Ms. Celia Lang who reduced a veritable mountain of rough drafts to a final typescript, and Mr. Karl Holeczek for his speedy and faithful execution of the figures and photography work.

So many people have shaped my thinking in this area of biophysics that it is difficult to acknowledge their individual contributions. The one person I can single out is Professor J. Walter Woodbury, who for twelve years has been a close friend, teacher, and colleague. Much of the material presented bears the imprint of his deft touch.

TABLE OF CONTENTS

PART I. INTRODUCTION

It is believed by most investigators (see Ling,1962, for the opposite opinion) that all biological cells are surrounded by a boundary layer that is chemically quite distinct from the extra-cellular fluid (ECF) and the intra-cellular fluid (ICF). This boundary layer which is called the cell, or plasma, membrane seems to perform many varied and specialized tasks necessary to cellular integrity and thus the integrity of whole multicellular organisms.

I am going to deal with only two of these specialized membrane functions, not because they are the most important (an impossible judgment to make) or the simplest to understand (unclear at this time), but rather because they afford an excellent opportunity to examine the way in which the biological sciences have been able to combine forces with the physical sciences and mathematics in gaining a deeper understanding of fundamental biological processes.

One broad area of membrane function that I examine theoretically is the way in which membranes can control the ionic composition of the ECF and ICF (see Table I.1). This is a basic area of interest for all types of cells, because it is known that many of the necessary intracellular chemical reactions in a viable cell are strongly dependent on the ionic medium in which they proceed. Further, the extracellular environment in which a cell exists is crucial for its continued survival and proper functioning. This basic problem will lead to an examination of the way in which ions move across biological membranes in response to a number of external forces, and the way in which they interact with the molecular

	Concentration (mM/Liter)			
	Squid Giant Axon		Frog Sartorius Muscle	
Substance	Blood	Axoplasm	Blood	Axoplasm
Na^+	335-440	32-50	145	12
K^+	20	400	2.5	155
Ca^{++}	10	0.4		
Mg^{++}	54	10		
Cl^-	560	40-150		

Table I.1. Concentration of inorganic ions in the axoplasm of the squid axon and frog sartorius muscle and the surrounding blood.

components of the membrane during their passage through it. It will also give some insight into how ionic movement through a membrane can contribute to the electrical properties of membranes as measured in experimental situations.

The second area of membrane function that I will look at theoretically incorporates all of the features of interest in the first group of phenomena with additional ones that are of importance in cells that must rapidly transmit information over fairly large distances. In lower organisms, information transfer between cells takes place primarily through the process of diffusion. Diffusion is a perfectly acceptable mode of information transfer as long as the time scales of importance are long with respect to the time necessary for diffusion (which, in turn, depends on the distance over which information must travel). Higher organisms are larger, and thus must relay information between cells that are spatially farther apart. Most of these organisms have evolved a system of communication that involves the rapid transmission of information by coded alterations in the electrical properties of the membranes of certain cells that collectively comprise the nervous system. It is the specialized membrane properties of cells in the nervous system that is the subject of our second area of investigation, and this will involve an examination of the mechanisms whereby cells generate and conduct action potentials (the basic information BIT in the nervous system).

CHAPTER 1. BASIC MEMBRANE STRUCTURE AND ELECTRICAL PROPERTIES

A. Some Simple Concepts About Membrane Structure. Theories related to the nature of membrane ultrastructure are legion but most of the interpretation of presently available data tends to favour the so called unit membrane hypothesis of Robertson (1957). Hendler (1971) has given a thorough review of the data pertinent to the structure of membranes, and the various theories that have been proposed to accomodate these data. He comes to the conclusion that, at the present, the unit membrane hypothesis is sufficient as an explanation. In what follows I have relied on Hendler's review in formulating the evidence for the unit membrane hypothesis.

The Unit Membrane Hypothesis (UMH). In Figure 1.1 the structure of a membrane according to the UMH is shown schematically, and should be referred to for what follows. The membrane is thin (75 $\pm$ 25 A) (1 A = 10^{-10}m) and composed of a central lipid core bounded

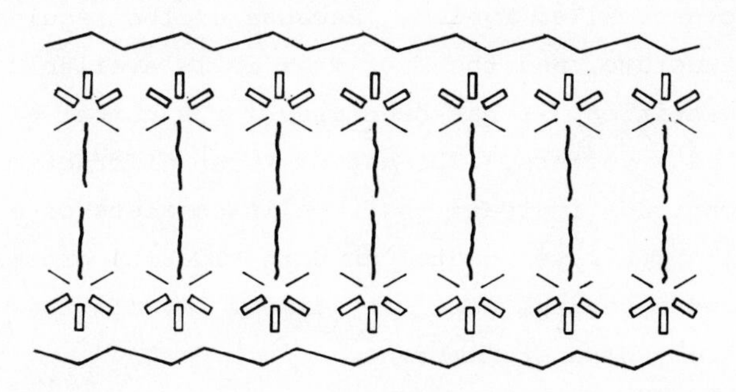

Figure 1.1 Membrane structure according to the UMH. denotes a lipid molecule; - is the polar head group, the hydrocarbon chain. is an extended β-configuration protein monolayer.

on both sides by monomolecular layers of protein and/or polysaccharides The lipid core is in a bilayer configuration with the (hydrophobic) hydrocarbon chains oriented toward the middle of the membrane and the (hydrophilic) polar head groups oriented toward the ECF and ICF. The monomolecular protein coats at each border are in the extended β configuration.

Evidence Supporting the UMH. The basic tenets of the UMH (composition and thickness) are not really in question. Overton (1895) established the fact that cellular membranes were composed, in part, of lipid by examining the rates of intracellular accumulation of a number of compounds with varying degrees of lipid solubility. He found that the higher the lipid solubility of a compound, the faster it entered cells . Thickness estimates for the plasma membrane are obtained from electronmicroscopy examination of natural cell membranes that have been fixed with $KMnO_4$. The results of these studies (see Robertson, 1957) consistently indicate the existence of a structure between 50 and 100 A thick.

It has been known for some time that lipid material, when spread on an aqueous phase, will form a mono-molecular film or monolayer (Bragg and Bragg, 1924; Langmuir, 1917). Gorter and Grendel (1925) extracted the lipids from the membranes of red blood cells, spread these in a monolayer, and observed that they occupied twice the estimated red blood cell surface area. This experiment, which has been repeated a number of times, supports the notion that the membrane lipids are arranged in a bilayer configuration.

Peripheral nerve fibres in many species are surrounded by a lipid-protein structure called myelin. Because of the regular nature of myelin structure, and the fact that is is available in relatively large quantities, it has been used for a number of analyses of its detailed structure. The use of X-ray diffraction and polarized light techniques indicate that myelin consists of a sheet like structure of lipid bilayer coated on both sides by protein that has been wound around a central core many times. Myelin thus seems to be composed entirely of unit membrane.

That the lipid bilayer is indeed an energetically stable structure was first demonstrated by Mueller et al. (1962) when they reconstructed membranes in the laboratory out of naturally occurring phospholipids. These artificial membranes are similar in many respects to natural membranes, including their appearance in electron micrographs, X-ray diffraction patterns, water permeability, and electrical capacitance. All of the available evidence from these laboratory contructed membrane analogs supports the concept that they are bilayer in nature. The most striking difference between artificial and naturally occurring membranes is that the artificial membrane has an electrical resistance many orders of magnitude larger than that found in natural membranes. A number of natural and synthetic polypeptides and proteins have been discovered that lower the artificial membrane resistance into the biological range. This, in conjunction with other evidence, leads to the conclusion that the ability of ions to move across biological membranes is due to the presence of very specific molecules, probably proteins, in the membrane.

B. The General Nature of Membrane Electrical Properties. Alternating current impedance measurements are one way of examining the electrical properties of the membrane in the resting state. Cole and Baker (1941a,b) made impedance measurements on the membrane of the squid giant axon and found that the impedance of the membrane behaved as if it were due to a capacitor and resistor in parallel Similar results have been found in the membrane of *Nitella* (Curtis and Cole, 1937).

The value of the membrane capacity, C_m, was found to be approximately 1 microfarad/cm^2 ($= 1\ \mu F/cm^2 = 10^{-6}\ F/cm^2$) of membrane for a variety of membranes from different species. If it is simply assumed that the membrane lipid acts as the dielectric medium (dielectric constant κ) for a parallel plate capacitor (spacing δ = membrane thickness), $C = \kappa\varepsilon_o/\delta$ implies that the membrane is of the order of 50 A in thickness. This is quite close to the value estimated from electron microscopy studies. This calculation is based on the assumption that the membrane dielectric constant is on the order of 5, not unreasonable for a structure composed primarily of lipid.

The membrane resistance R_m was found to have a value on the order of 10^3 to 10^4 ohm-cm^2 of membrane. This resistance most certainly represents the portion of the membrane through which ions move. Considerations of the factors determining R_m will be of prime concern in the material that follows because of its importance for cellular homeostasis.

It is possible to construct microelectrodes with a tip diameter on the order of 0.5 microns and introduce them inside biological cells (Ling and Gerard, 1949). When this is done it is found that there is an electrical potential difference (V_m) across the membrane on the order of 100 mV (1 millivolt = 1 mV = 10^{-3} V), with the inside of the cell being negative with respect to the outside (measurements of this potential, the membrane potential, will always be referred to the outside of the cell). From the above estimates of membrane thickness, and assuming that the membrane potential varies in a linear fashion across the membrane, the electrical field strength, E, within the membrane ($E = -V_m/\delta$) is on the order of 10^5 volts/cm. Thus in a resting state biological membranes operate in the presence of very strong fields that are close to the fields required to cause dielectric breakdown in materials with a $\kappa \sim 5$.

The existence of potential differences and ionic concentration differences across the membrane lead to the conclusion that generally the transmembrane electrochemical energy difference (ΔU) for an ion at concentration n_i (number/cm^3) in the ICF and n_o in the ECF will not be zero:

$$\Delta U = kT \ln(n_i/n_o) + ZeV_m$$

wherein k is Boltzmanns constant, T is the temperature in $^\circ$K, Z is the valence of the ion, and e is electronic charge. When $\Delta U = 0$, the ion in question is said to be in electrochemical equilibrium, and the condition that must attain in that situation is $V_m = V_e = -(kT/Ze)\ln(n_i/n_o)$ where V_e is the equilibrium potential.

When current is passed through the squid axon membrane (Cole and Baker, 1941a) there is a decrease in the resistance of the parallel membrane equivalent circuit for an inward current, and almost no change in the reactance (capacitance). When outward current is passed, the parallel resistance increases and there is a small reactance change.

Hodgkin and Rushton (1946) made more precise measurements of the membrane resistance and capacitance than had been possible in the impedance experiments. They proposed, and theoretically developed, a model for the spatial and temporal spread of currents in a single axon. In formulating the model it was assumed that the membrane resistance was ohmic, valid for small membrane currents, the membrane capacitance was ideal, and that current flow in the axoplasm and interstitial fluid were parallel. From a set of experiments on the lobster giant axon they concluded that the membrane resistance was 600-700 ohm-cm^2, and that the membrane capacity was 1.3 μf/cm^2.

The membrane properties discussed above were deduced from electrical measurements. Others have been discovered by ionic flux measurements. Hodgkin and Keynes (1955) made measurements of the influx and efflux of radioactive Na^+ and K^+ in the axons of squid and cuttlefish, and the effect of certain metabolic posions on the fluxes. It was found that the movement of Na^+ and K^+ down their respective electrochemical gradients (i.e., Na^+ influx and K^+ efflux) is relatively unaffected by the action of such metabolic inhibitors as dinitrophenol (DNP), cyanide, and sodium azide, or by cooling to 1°C. Quite a different situation is found however, when the movement of these ions against their electrochemical gradients is examined under the effects of cooling or inhibitors. Addition of DNP or cyanide to the interstitial fluid has the reversible effect of greatly reducing K^+ influx and Na^+ efflux as does cooling. It is significant that the decrease in K^+ uptake is approximately equal to the decrease in Na^+ extrusion. If $n_{K_o} = 0$, it is further found that there is a reduction by one third in the Na^+ efflux. As a result of findings such as these the existence of a mechanism (Na^+-K^+ 'pump') capable of doing work against the electrochemical gradient in the membrane to extrude Na^+ and take up K^+ was confirmed. It is hypothesized that this pump is driven by metabolic energy and is effectively stopped (reversibly) by cooling or metabolic inhibitors. Hodgkin and Keynes further demonstrated that poisoning of this pump in no way directly affects the impulse conduction, but does prevent recovery.

In a second paper, Hodgkin and Keynes (1955b) investigated the permeability of the membrane to potassium. Using labeled K^+ in single fibers poisoned by DNP, they found that the net K^+ flux across the membrane is in the direction of its electrochemical gradient, and that the net flux is reduced to zero when the membrane potential is equal to the K^+ equilibrium potential. These two observations lead to the conclusion that K^+ transport across a poisoned membrane is due to passive factors only. By further measuring K^+ fluxes as a function of V_m, it was found that the permeability (and hence the conductance) of the membrane to K^+ is greatly elevated when the membrane is depolarized.

During excitation (a term which collectively refers to a complex series of membrane related events) there is a transient change in the cell membrane potential from its resting level of some -80 to -100 mV to about +40 mV and then back to its resting value (see Figure 1.2). This transient change, generally lasting about 1 msec,

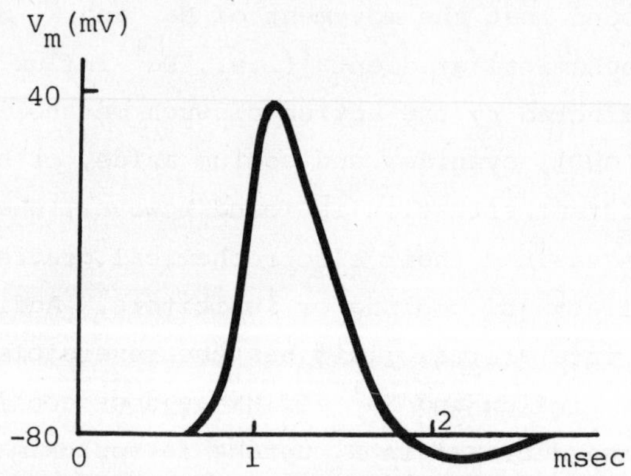

Figure 1.2. The form of the propogating action potential as a function of time at a fixed location on the axon.

is called the action potential. The action potential travels at a constant velocity and with fixed shape along the cell axon. Cole and Curtis (1938, 1939) conclusively demonstrated that the action potential is associated with a transient decrease in membrane resistance from approximately 1000 Ω-cm^2 at rest to 10 Ω-cm^2 at the peak of activity.

PROBLEMS

1.1. A recently isolated pure protein fraction from locust flight muscle (LTD for short) has a molecular weight of 600,000. The addition of 1 nanogram of LTD/cm^2 of artificial membrane lowers the specific membrane resistance from 10^9 ohm-cm^2 to 10 ohm-cm^2 when the membrane is bathed by 10^{-3} M/L KCl on each side. (1 nanogram = 10^{-9} grams).

a. How many LTD molecules are there in one square micron of membrane (1 micron = 10^{-6} metre)?

b. Assuming that one LTD molecule forms a "channel" through which ions may pass, what is the resistance of one channel?

c. Assuming that each channel is "filled" with 10^{-3} M/L KCl (resistivity, ρ, is 10 ohm-cm) what is the equivalent cross sectional area of one channel?

1.2. From the data of Table 1.1 calculate the equilibrium potentials of Na^+, K^+, Cl^-, Ca^{++}, and Mg^{++} for the squid giant axon and for a frog sartorius muscle

$R = \frac{\rho l}{a}$

CHAPTER 2. PASSIVE ELECTRICAL PROPERTIES OF AXONS

To obtain a feeling for the difficulties that the central and peripheral nervous system would face in transmitting information encoded as changes in the membrane potential of neurons without the action potential, it is instructive to explore the cable properties of long cells (i.e., axons). In this chapter I first examine the steady state and time dependent membrane potential responses of a non-myelinated fibre to injected current, and then examine the analogous steady state problem for a myelinated fibre.

Further work on cable properties may be found in Cole and Hodgkin (1939), Hodgkin and Rushton (1946), Tasaki (1953) and Taylor (1963). For examples of the use of cable theory in investigating the role played by the dendritic tree in synaptic integration, reference should be made to Rall (1959, 1960, 1962a,b, 1964, 1967) and Butz and Cowan (1974).

A. Non-Myelinated Fibres. As illustrated in Figure 2.1, if a current I_o is injected into an unmyelinated cell it will spread in

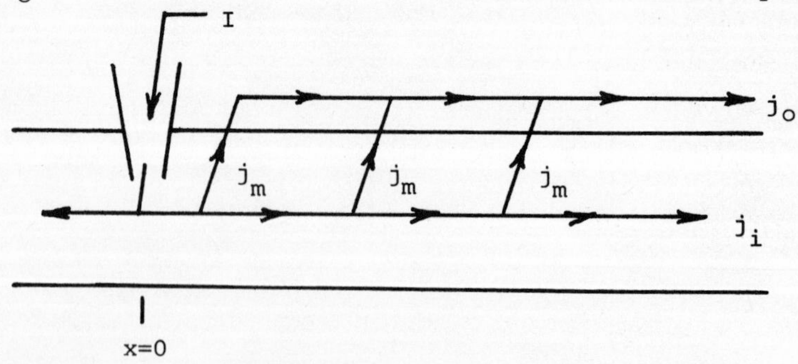

Figure 2.1 Spread of injected current, I, in an unmyelinated axon.

both directions before passing out through the membrane. Let the current along the interior of the cell be j_i, through the membrane j_m, and the current in the extracellular fluid j_o. The electrical properties of the membrane are characterized by the membrane resistance R_m (ohm-cm^2) and capacity C_m (μF/cm^2), and the axoplasm by a specific resistivity ρ_i (ohm-cm).

An alternate way of characterizing these axonal properties is to consider a unit length (e.g., 1 cm) of axon. Thus if the axon is of radius a then $r_m = R_m/2\pi a$ (Ω-cm), $c_m = 2\pi a C_m$ (μF/cm), and $r_i = \rho_i/\pi a^2$ (Ω/cm). If the axon is thought of as being cut into cylindrical segments of length Δx then these definitions can be used to draw the axonal equivalent circuit as shown in Figure 2.2. r_o is the ECF resistance per unit length, and the batteries represent the membrane resting potential V_R.

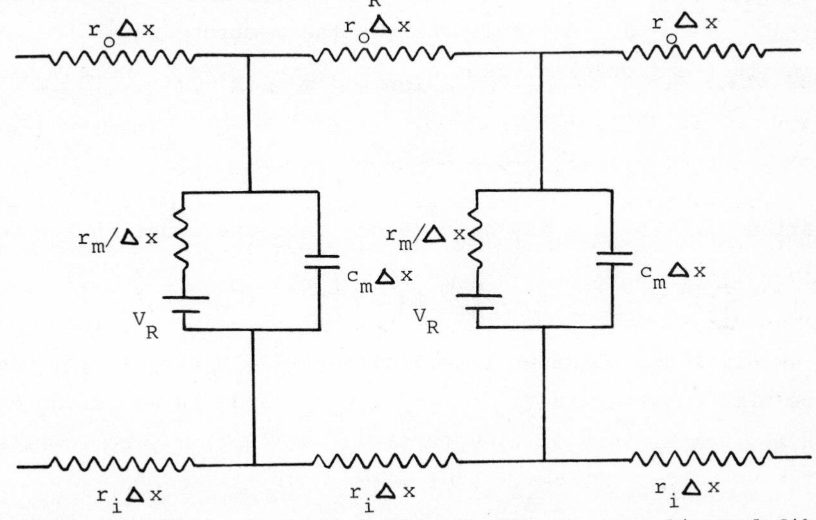

Figure 2.2 Equivalent electrical circuit for an unmyelinated fibre when V_m is near V_R.

If the subscripts "o", "i", and "m" denote exterior, interior, and membrane variables respectively, then ohms law applied to the extracellular fluid and axoplasm gives

$$\frac{\partial V_o}{\partial x} = -r_o j_o \tag{2.1}$$

and

$$\frac{\partial V_i}{\partial x} = -r_i j_i \tag{2.2}$$

Further, the total longitudinal current flow is $j = j_o + j_i$ and the membrane potential is $V_m = V_i - V_o$. By the continuity of current (Kirchoff's Law) the membrane current j_m (per unit length of axon) is given by

$$j_m = -\frac{\partial j_i}{\partial x} = \frac{\partial j_o}{\partial x} \tag{2.3}$$

Subtracting (2.1) from (2.2)

$$\begin{aligned}\frac{\partial V_i}{\partial x} - \frac{\partial V_o}{\partial x} = \frac{\partial V_m}{\partial x} &= -r_i j_i + r_o j_o \\ &= -(r_i + r_o) j_i + r_o (j_i + j_o) \\ &= -(r_i + r_o) j_i + r_o j\end{aligned} \tag{2.4}$$

Differentiating (2.4) and using the fact that $(\partial j/\partial x) = 0$ by (2.3) gives

$$\frac{\partial^2 V_m}{\partial x^2} = -(r_i + r_o)\frac{\partial j_i}{\partial x} = (r_i + r_o) j_m \tag{2.5}$$

The membrane current has two components; one is the displacement current $j_D = c_m(\partial V_m/\partial t)$ that charges the membrane capacitor and the second is the (ionic) current $j_{ion} = (V_m - V_R)/r_m$, where V_R is the resting potential. Thus

$$j_m = c_m \frac{\partial V_m}{\partial t} + \frac{V_m - V_R}{r_m} \tag{2.6}$$

Equations 2.5 and 2.6 may be combined into one equation for V_m:

$$\frac{\partial^2 V_m}{\partial x^2} = c_m(r_i + r_o)\frac{\partial V_m}{\partial t} + \left(\frac{r_i + r_o}{r_m}\right)(V_m - V_R) \tag{2.7}$$

The axonal space constant is defined by $\lambda^2 = r_m/(r_o + r_i)$, the membrane time constant is $\tau_m = r_m c_m = R_m C_m$, and the deviation of $V_m(x,t)$ from V_R is $\bar{V}_m(x,t) = V_m(x,t) - V_R$, so (2.7) may be rewritten as

$$\lambda^2 \frac{\partial^2 \bar{V}_m}{\partial x^2} - \bar{V}_m = \tau_m \frac{\partial \bar{V}_m}{\partial t} \tag{2.8}$$

Equation 2.7, or 2.8, is the cable equation and was used by Lord Kelvin to theoretically treat signal transmission on trans-Atlantic submarine cables. Further discussions of passive axonal electrical properties will be based on solutions of (2.8), and the significance of the parameters λ and τ_m will become clear through this discussion.

In a steady state the current supplied to the axon via the microelectrode has been on for a sufficiently long time so $\bar{V}_m$ has reached a steady value. Thus take $(\partial \bar{V}_m/\partial t) = 0$, which will be true when the membrane capacity is charged and no displacement current is flowing. Equation 2.8 becomes, with $\lim_{t \to \infty} \bar{V}_m(x,t) = \bar{V}_{mss}$

$$\lambda^2 \frac{d^2\bar{V}_{mss}}{dx^2} = \bar{V}_{mss} \tag{2.9}$$

which has the general solution

$$\bar{V}_{mss} = A_1 \exp(x/\lambda) + A_2 \exp(-x/\lambda) \tag{2.10}$$

To determine the constants in (2.10) requires the specification of boundary conditions on $\bar{V}_{mss}(x)$. For this particular problem as $x \to \pm\infty$, $\bar{V}_{mss}$ must remain finite. It is further required that $\bar{V}_{mss}$ be continuous at $x = 0$. (If $\bar{V}_{mss}$ were not continuous at $x = 0$ it would imply the existence of some mechanism within the axon capable of giving rise to an unequal current flow on each side of $x = 0$). These criteria applied to (2.10) give, with $A_1 = A_2 = A$,

$$\bar{V}_{mss} = A\exp(-|x|/\lambda) \tag{2.11}$$

The constant A in (2.11) may be determined from the current I_o injected at $x = 0$. It is clear that

$$I_o = \int_{-\infty}^{\infty} j_m(x)\,dx = 2\int_0^{\infty} j_m(x)\,dx \tag{2.12}$$

by definition and symmetry. From equations 2.5 and 2.11

$$I_o = \frac{2}{r_o+r_i}\int_0^{\infty} \frac{\partial^2\bar{V}_m}{\partial x^2}\,dx = \frac{2A}{\lambda(r_o+r_i)}$$

so

$$A = \frac{I_o\lambda(r_o + r_i)}{2} \tag{2.13}$$

and therefore

$$\bar{V}_{mss}(x) = \frac{(r_o+r_i)\lambda I_o}{2}\exp(-|x|/\lambda) \tag{2.14}$$

The input resistance, R_c, of an axon is operationally defined as the ratio of the steady-state voltage change at $x = 0$, $\bar{V}_{mss}(0)$, to the current that produced it, I_o, so

$$R_c = \frac{(r_i + r_o)\lambda}{2}$$

by (2.14). Thus (2.14) may be rewritten as

$$\bar{V}_{mss}(x) = I_o R_c \exp(-|x|/\lambda) \tag{2.15}$$

a form of Ohm's Law.

The significance of λ, the space constant, is now clear from equation 2.14 or 2.15. It is simply the distance at which $\bar{V}_{mss}(x)$ falls to e^{-1} times its value at $x = 0$. This quantity, like the input resistance, is easily obtained from data taken in an experiment. Typical values of λ fall in the range 2-10 mm, thus indicating that the simple axonal model treated so far is incapable of transmitting steady state information over any appreciable distance.

To examine the solution of the cable equation when V_m is allowed to vary with respect to both x and t in response to a step change in current at (x = 0, t = 0), make two transformations of the independent and dependent variables in (2.8). First define two new dimensionless independent variables by $X = x/\lambda$ and $T = t/\tau$. With these new variables, (2.8) may be written

$$\frac{\partial^2 \bar{V}_m}{\partial X^2} - \bar{V}_m = \frac{\partial \bar{V}_m}{\partial T} \tag{2.16}$$

Secondly, transform the dependent variable by

$$U = \bar{V}_m \exp T \tag{2.17}$$

Thus $V_m = U \exp(-T)$, $(\partial^2 \bar{V}_m / \partial X^2) = (\partial^2 U / \partial X^2)\exp(-T)$, and $(\partial \bar{V}_m / \partial T) = [(\partial U / \partial T) - U]\exp(-T)$ so (2.16) becomes

$$\frac{\partial^2 U}{\partial X^2} = \frac{\partial U}{\partial T} \tag{2.18}$$

To solve (2.16) or (2.18) boundary, initial and continuity conditions must be specified. These are: $\lim_{x \to \infty} \bar{V}_m(x,t) = 0$; $\bar{V}_m(x,t) = 0$ for $-\infty < t < 0$; and $\bar{V}_m(x,t)$ is continuous at $x = 0$ respectively.

Take the Laplace transform of equation 2.18 with respect to T, giving

$$\begin{aligned} \frac{d^2 u}{dX^2} &= su - U(T{=}0) \\ &= su \end{aligned} \tag{2.19}$$

by the initial condition, where $\mathcal{L}[U(X,T)] = u(X,s)$. The solutions of (2.19) may be written

$$u(X,s) = \begin{cases} A_1 \exp(\sqrt{s}\,X) + A_2 \exp(-\sqrt{s}\,X) & X \leq 0 \\ B_1 \exp(\sqrt{s}\,X) + B_2 \exp(-\sqrt{s}\,X) & X \geq 0 \end{cases} \tag{2.20}$$

The boundary conditions imply that $A_2 = B_1 = 0$ while the continuity condition implies $A_1 = B_2 = A$. Thus (2.20) reduces to

$$u(X,s) = A\exp(-\sqrt{s}\,X) \tag{2.21}$$

To determine the constant A proceed as before, giving

$$I_o = 2\int_0^\infty j_m(x)dx = \frac{2}{r_i+r_o}\int_0^\infty \frac{\partial^2 \bar{V}_m}{\partial x^2}dx \tag{2.22}$$

With the definition of X this may be rewritten as

$$I_o = \frac{2}{\lambda(r_i+r_o)}\int_0^\infty \frac{\partial^2 \bar{V}_m}{\partial X^2}dX \tag{2.23}$$

Multiply through in (2.23) by exp(T) so

$$I_o \exp T = \frac{2}{\lambda(r_i+r_o)}\int_0^\infty \frac{\partial^2 U}{\partial X^2}dX \tag{2.24}$$

Taking the Laplace transform of (2.24) and using (2.19) and (2.21) gives

$$I_o = \frac{2A\sqrt{s}\,(s-1)}{\lambda(r_o+r_i)}$$

or

$$A = \frac{I_o\lambda(r_i+r_o)}{2}\cdot\frac{1}{\sqrt{s}\,(s-1)}$$

and the Laplace transform of the solution may be written

$$u(X,s) = \frac{\lambda I_o(r_i+r_o)}{2}\cdot\frac{\exp(-\sqrt{s}\,|X|)}{\sqrt{s}\,(s-1)} \tag{2.25}$$

The last thing that remains is to take the inverse Laplace transform of (2.25) to obtain an expression for $\bar{V}_m(X,T)$. Consider the solution for $X>0$ only as the solution for $X<0$ is easily obtained.

Writing the $X>0$ portion of (2.25) in partial fractions gives

$$u(X,s) = \frac{\lambda I_o(r_i+r_o)}{2}\exp(-\sqrt{s}\,X)\left\{\frac{1}{2}\left[\frac{1}{\sqrt{s}-1}+\frac{1}{\sqrt{s}+1}\right]-\frac{1}{\sqrt{s}}\right\} \tag{2.26}$$

Now (Roberts and Kaufman, 1966),

$$\mathcal{L}^{-1}\left[\frac{\exp(-\sqrt{s}\,X)}{\sqrt{s}}\right] = \frac{1}{\sqrt{\pi T}}\exp(-X^2/4T),$$

$$\mathcal{L}^{-1}\left[\frac{\exp(-\sqrt{s}\,X)}{\sqrt{s}\mp 1}\right] = \frac{\exp(-X^2/4T)}{\sqrt{\pi T}} + \left\{1-\operatorname{erf}\left[\frac{X}{2\sqrt{T}}\mp\sqrt{T}\right]\right\}\exp(T\mp X)$$

where the error function, erf(z), is defined by

$$\operatorname{erf} Z = \frac{2}{\sqrt{\pi}}\int_0^Z \exp(-w^2)dw$$

Taking the inverse transform of (2.26) in conjunction with the above,

$$U(X,T) = \frac{I_o\lambda(r_i+r_o)}{2}\left\{\left[1-\text{erf}\left(\frac{X}{2\sqrt{T}} - \sqrt{T}\right)\right]\exp(T-X) - \left[1-\text{erf}\left(\frac{X}{2\sqrt{T}} + \sqrt{T}\right)\right]\exp(T+X)\right\} \qquad (2.27)$$

or with $U = \bar{V}_m \exp T$

$$\bar{V}_m(X,T) = \frac{I_o\lambda(r_i+r_o)}{2}\left\{\left[1-\text{erf}\left(\frac{X}{2\sqrt{T}} - \sqrt{T}\right)\right]\exp(-X) - \left[1-\text{erf}\left(\frac{X}{2\sqrt{T}} + \sqrt{T}\right)\right]\exp X\right\} \qquad (2.28)$$

Using (2.28) Hodgkin and Rushton (1946) were able to fit their passive electrical data in crustacean axon with R_m = 600 ohm-cm^2, C_m = 1.3μF/cm^2. At a given X, the variation of $\bar{V}_m(X,T)$ with T is as found experimentally, rising faster than a simple exponential of the form $(1 - e^{-T})$ which would be expected in the absence of cable properties.

B. Myelinated Fibres. A myelinated nerve looks like the representation in Figure 2.3. The axon has a surrounding lipid (myelin) region that is interrupted at regular intervals by regions called nodes of Ranvier. All available evidence indicates that excitation occurs only at the nodes, and that excitation "skips" from node to node (saltatory conduction). For a single fibre from the frog sciatic nerve, the data of Table 2.1 is relevant. With the above comments and data in mind I will model the myelinated nerve by the electrical equivalent circuit of Figure 2.4.

Note that I am using a discrete approximation for the electrical structure, rather than the continuous representation of our non-myelinated axon consideration. In this discrete approximation the internodal portions of the axon are represented as lumped parameters. A more precise, and difficult, means of attack would take the continuous spatial characteristics of the internode into account (c.f. Pickard, 1969). Instead of membrane potential as a function of position (x) along the nerve and time (t) in response to an applied current at x = 0, I want V_m as a function of node number (n) and time where node n = 0 is the site of current injection.

With the same notation as before for voltages and currents,

$$V_o(n+1,t) - V_o(n,t) = -r_o j_o(n,t) \qquad (2.29)$$

gives the difference in outside potential, $V_o(n,t)$ between two

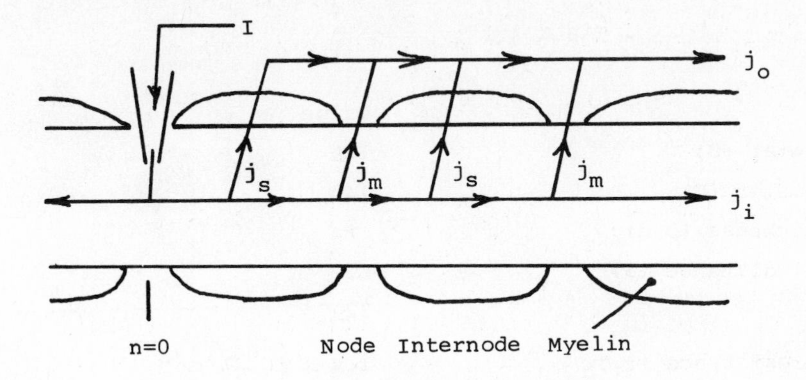

Figure 2.3 Schematic representation of the structure, and spread of injected current, in a myelinated fibre.

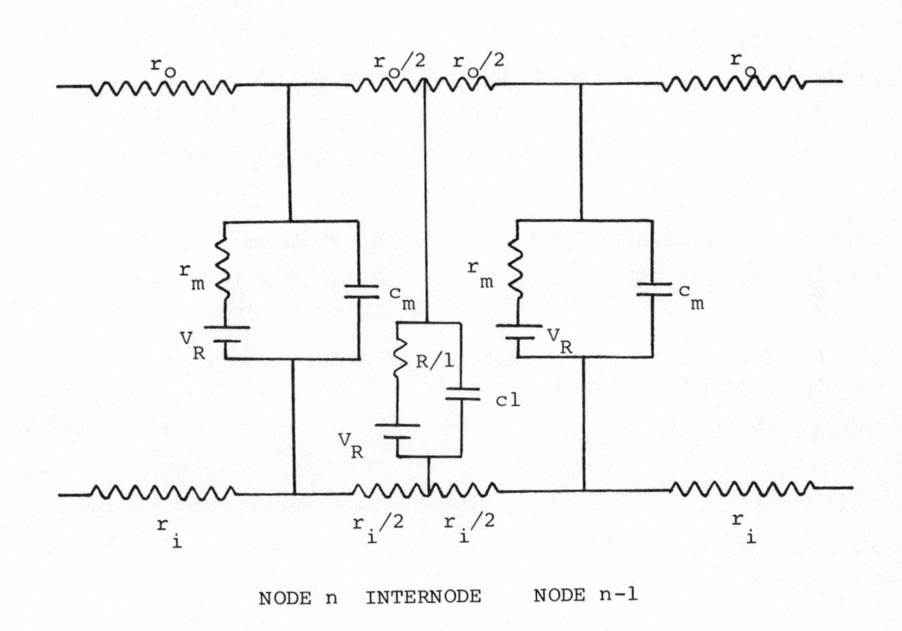

Figure 2.4 Lumped equivalent electrical circuit for a myelinated fibre.

TABLE 2.1

Parameter	Value
Axon diameter (d)	12μ
Fibre diameter (D)	14μ
Myelin thickness (D-d)	2μ
Internodal distance (l)	0.2 cm
Nodal area	$22\mu^2$
Axoplasm resistance (r_i)	$1.4 \times 10^8 \Omega$/cm
Axoplasm resistivity (ρ_i)	$1.1 \times 10^2 \Omega$/cm
Myelin capacity (c)	10^{-11} F/cm
Specific myelin capacity (C)	2.5×10^{-9} F/cm
Myelin resistivity (ρ)	$8 \times 10^8 \Omega$-cm
Specific myelin resistance (R)	$1.6 \times 10^5 \Omega\text{-cm}^2$
Myelin time constant (τ=RC)	4.0×10^{-4} sec
Internode time constant ($\tau_i = r_i c l^2$)	5.6×10^{-5} sec
Nodal capacity (c_m)	0.6×10^{-13} F
Specific nodal capacity (C_m)	3.0×10^{-6} F/cm^2
Nodal resistance (r_m)	$4\text{-}8 \times 10^7 \Omega$
Specific nodal resistance (R_m)	$10\text{-}20 \Omega\text{-cm}^2$
Nodal time constant ($\tau_m = R_m d_m$)	$2.4\text{-}4.8 \times 10^{-5}$ sec

Table 2.1. Typical fibre parameters for a frog myelinated fibre. From Stämpfli (1952).

adjacent nodes, and

$$V_i(n+1,t) - V_i(n,t) = -r_i j_i(n,t) \tag{2.30}$$

gives the difference in inside potential, $V_i(n,t)$. r_o and r_i are now for one internodal length, rather than one cm of axon. Defining the membrane potential at node n by

$$V_m = V_i(n,t) - V_o(n,t)$$

and the total current flowing in an internode by

$$j(n,t) = j_o(n,t) + j_i(n,t)$$

equations 2.29 and 2.30 become

$$V_m(n+1,t) - V_m(n,t) = -(r_i+r_o)j_i(n,t) + r_o j(n,t) \tag{2.31}$$

Further, between node (n+2) and (n+1)

$$V_m(n+2,t) - V_m(n+1,t) = -(r_i+r_o)j_i(n+1,t) + r_o j(n+1,t) \tag{2.32}$$

so (2.31) and (2.32) combine to yield

$$V_m(n+2,t) - 2V_m(n+1,t) + V_m(n,t) = (r_i+r_o)\left[j_i(n,t) - j_i(n+1,t)\right] + r_o\left[j(n+1,t) - j(n,t)\right] \tag{2.33}$$

The current through the nth node (j_m) and the nth internode (j_s) are related to the intra- and extracellular currents by

$$\begin{aligned} j_s(n,t) + j_m(n,t) &= j_o(n,t) - j_o(n-1,t) \\ &= j_i(n-1,t) - j_i(n,t) \end{aligned} \tag{2.34}$$

which implies that

$$j_i(n,t) + j_o(n,t) = j_i(n-1,t) + j_o(n-1,t)$$

or

$$j(n,t) = j(n-1,t) \tag{2.35}$$

With (2.34) and (2.35), (2.33) may be rewritten as

$$\begin{aligned} &V_m(n+2,t) - 2V_m(n+1,t) + V_m(n,t) \\ &= (r_i+r_o)\left[j_m(n+1,t) + j_s(n+1,t)\right] \end{aligned} \tag{2.36}$$

This is a difference equation relating the membrane potential at nodes (n+2), (n+1), and n to the nodal current at node (n+1) and the sheath current in internode (n+1). Both of these currents are sums of ohmic and displacement components:

$$j_m(n,t) = c_m \frac{\partial V_m(n+1,t)}{\partial t} + \frac{V_m(n+1,t)}{r_s} \tag{2.37}$$

and

$$j_s(n,t) = c_s \frac{\partial V_m(n+1,t)}{\partial t} + \frac{V_m(n+1,t)}{r_s}$$

Combining (2.36) and (2.37) yields a second order difference equation with only one dependent variable, $V_m(n,t)$, to solve for:

$$V_m(n+2,t) - 2V_m(n+1,t) + V_m(n,t)$$

$$= (r_i+r_o)\left[(c_m+c_s)\frac{\partial V_m(n+1,t)}{\partial t} + \frac{r_m+r_s}{r_m r_s}V_m(n+1,t)\right] \qquad (2.38)$$

Here I will only examine the steady state solution of (2.38). One time dependent solution will be explored in Chapter 3 with respect to conduction velocity estimates.

In a steady state situation, equation 2.38 becomes

$$V_m(n+2) - \left[2 + \frac{(r_i+r_o)(r_m+r_s)}{r_m r_s}\right]V_m(n+1) + V_m(n) = 0 \qquad (2.39)$$

In a fashion quite similar to that employed for solving differential equations, try a solution of the form

$$V_m(n) = Ap^n \qquad (2.40)$$

where A and p are constants to see that it is indeed a solution of (2.39) if and only if

$$p^2 - \left[2 + \frac{(r_i+r_o)(r_m+r_s)}{r_m r_s}\right]p + 1 = 0 \qquad (2.41)$$

Clearly from (2.41) there are two values of p, p_1 and p_2, which make (2.40) a valid solution of (2.39). Thus, quite generally,

$$V_m(n) = A_1 p_1^{\,n} + A_2 p_2^{\,n} \qquad (2.42)$$

From the coefficients of (2.41) it is clear that $p_1 p_2 = 1$ and

$$p_1 + p_2 = 2 + \frac{(r_i+r_o)(r_m+r_s)}{r_m r_s}$$

Instead of dealing with the p's it is instructive to define a new variable Γ by

$$p_1 = \exp(-\Gamma) \qquad (2.43)$$

so

$$p_2 = p_1^{-1} = \exp(\Gamma)$$

and

$$(p_1 + p_2)/2 = \cosh \Gamma \qquad (2.44)$$

Thus I may rewrite equation 2.42 as

$$V_m(n) = A_1\exp(-n\Gamma) + A_2\exp(n\Gamma) \tag{2.45}$$

To evaluate the constants A_1 and A_2 in (2.45) boundary conditions are required. First, $V_m(n)$ must approach zero as n becomes large. Thus $A_2 = 0$ for n positive, and (2.45) becomes (with $A = A_1$)

$$V_m(n) = A\exp(-n\Gamma) \tag{2.46}$$

To evaluate the constant A, proceed much as in the non-myelinated case. From equation 2.36

$$j_s(n) + j_m(n) = (r_i+r_o)^{-1}\left[V_m(n+1) - 2V_m(n) + V_m(n-1)\right]$$

so, in combination with (2.46),

$$j_s(n) + j_m(n) = \frac{A\exp(-n\Gamma)}{r_i + r_o}\left[\exp(-\Gamma) + \exp\Gamma - 2\right] \tag{2.47}$$

If a total current I_o is injected at n = 0, then

$$I_o = -j_m(o) + 2\sum_{n=o}^{\infty}\left[j_m(n) + j_s(n)\right]$$

and this, in conjunction with (2.47) gives

$$I_o = A(r_i+r_o)^{-1}\left[\exp\Gamma + \exp(-\Gamma) - 2\right]\frac{1 + \exp(-\Gamma)}{1 - \exp(-\Gamma)}$$

where I used the relation

$$\sum_{K=0}^{\infty} a^{Kx} = (1-a^x)^{-1}$$

Now from (2.43) and (2.44)

$$\exp\Gamma + \exp(-\Gamma) - 2 = (r_i+r_o)(r_m+r_s)/r_m r_s$$

and

$$\frac{1 + \exp(-\Gamma)}{1 - \exp(-\Gamma)} = \coth(\Gamma/2)$$

so

$$I_o = \frac{A}{r_m}\coth(\Gamma/2)$$

or

$$A = I_o r_m \tanh(\Gamma/2)$$

and thus

$$V_m(n) = I_o r_m \tanh(\Gamma/2)\exp(-n\Gamma)$$

As in the non-myelinated axon, I define the input resistance as $R_c = V_m(0)/I_o = r_m\tanh(\Gamma/2)$ so $V_m(n) = I_o R_c\exp(-n\Gamma)$. The constant Γ^{-1} in this steady state treatment of myelinated nerve properties plays the role of the space constant λ from the non-myelinated case. For the 14μ fibre of Table 2.1 a value of $\Gamma^{-1} = 0.5$ internode is predicted, and this corresponds to 10 mm for the 0.2 cm internode region.

The results of these calculations clearly illustrate that the simple models for the myelinated or unmyelinated axon that have been analyzed with respect to their cable properties are inadequate to explain how information can be transmitted over any appreciable distance, e.g. up to 2 metres in man. As will be seen later, the flaw is in the choice of a constant, field independent membrane resistance R_m. Actually R_m is a highly nonlinear function of time and membrane potential. However the above analyses are valid for small changes in the membrane potential around its resting value, and will offer valuable insight into the role of threshold effects and axonal parameters in determining conduction velocity (Chapter 3).

PROBLEMS

2.1. Ultimately it is necessary to determine the values of the parameters that characterize an axon electrically, e.g. R_m and C_m. Compare and contrast the kinds of information available from steady state and time dependent measurements on both myelinated and nonmyelinated fibres. How would you utilize these data to determine axonal parameters? Be sure to illustrate your discussion with diagrams whenever appropriate.

2.2. For the myelinated fibre can the expression for $\bar{V}_m(X,T)$ in response to a constant current turned off at T = 0 be obtained from the analogous expression for current turned on at T = 0 merely by changing the sign of T? Justify your conclusion.

2.3. Use the superposition principle (if valid) or some other method to calculate the membrane voltage response to the break of current at T = 0 for a nonmyelinated axon.

2.4. In the text I derived membrane potential response expressions for the myelinated axon only when $X > 0$. Derive the analogous expressions for $X < 0$.

2.5. Demonstrate that $\lim_{T \to \infty} \bar{V}_m(X,T)$ (equation 2.28) $= \bar{V}_{mss}(X)$.

2.6. Derive an expression for the passive membrane potential response of an unmyelinated axon to a current pulse of height I_o that is turned on at T = 0 and turned off at T = A. Using your result, graph the response at X = 1,2, and 3 for a pulse with A = 1.

2.7. At X = 0 in an unmyelinated axon a current $I_o \exp(j W T)$, where $W = \omega \tau_m$ is a dimensionless angular frequency, is injected via an intracellular microelectrode. Find $\bar{V}_m(X,T,W)$, and display a formula for $\bar{V}_{mss}(X,W)$. For X = 0, plot the imaginary part of $Z^*(X,W) = \bar{V}_{mss}(X,W)/I_o \exp(jWT)$ against the real part, with W as a parametric variable, and discuss the resulting diagram. What role do the cable properties of the fibre play in determining the shape of the curve? (i.e., repeat the calculation and graphical analysis for $r_o = r_i = 0$).

CHAPTER 3. OVERVIEW OF THE GROSS PROPERTIES OF EXCITABLE CELLS

A. The membrane theory of excitation and propagation. Over the past 70 years there has been an accumulation of data on excitable membrane systems leading to the formulation of the "membrane theory of excitation". The membrane theory of today is much different from, but based on, that of Bernstein (1902). No attempt will be made to trace the development of the membrane theory to its present form since Cole (1968) has already done so. Rather I will present the salient points of the theory together with the major supporting evidence. For more complete documentation of these points, reference should be made to reviews by Hodgkin (1951, 1958, 1964) and Cole (1968).

Foundations of the membrane theory.

1. There exists a thin, poorly conducting membrane surrounding the axon which divides plasma from interstitial fluid and is the seat of the resting and excitable properties of the axon.

2. In the resting state the axon membrane is more permeable to cations than anions, and more permeable to potassium than any other cation.

3. During excitation the membrane transiently becomes more permeable to sodium than to potassium.

4. Propagation of the action potential along the axon is by means of local circuit current flow between resting and excited regions.

The first statement, the central one of the theory, implies

that the resting and excited properties of the axon should be relatively independent of the axoplasm. Successful perfusion experiments by many investigators after the initial ones of Baker, Hodgkin, and Shaw (1961) lend adequate support to this conclusion. The squid giant axon functions normally when the axoplasm is replaced with artificial solutions, indicating that the molecular mechanisms responsible for the excitable properties of the axon reside within or are closely attached to the membrane. Another consequence of the first statement is that destruction of the membrane should destroy the resting and excitable properties of the axon. As discussed earlier and in Woodbury _et al_.(1969) the membrane probably has as its major constituents lipid and protein. Narahashi and Tobias (1964), Takenada and Yamagishi (1966), Albuquerque and Thesleff (1967), Gainer (1967) and Condrea, Rosenberg and Dettbarn (1967) have all shown that lipases and proteases destroy the resting potential and action potential in both single axons and muscle fibers. The result of these treatments has been the partial or total destruction of the surface membrane of the axon.

The most compelling evidence for the thinness of the membrane comes from the capacitance measurements of Fricke (1925) on red blood cells. He deduced that these cells had a surface membrane with a capacity of 0.8 $\mu F/cm^2$. Previous experiments indicated that this membrane was primarily lipid. Thus the static dielectric constant is about 3 and from the parallel plate capacitor formula the membrane thickness is about 33 Å. This figure has been obtained by a number of investigators, and the membrane thickness is generally considered to be on the order of 50-100 A.

Specific membrane resistances usually fall in the range 10 to 10^4 ohm-cm^2. For a membrane thickness of 100 A this places the membrane resistivity in the range 10^7 to 10^{10} ohm-cm, comparable with values for semiconductors. This is in sharp contrast to the electrolyte bathing the membrane with resistivities of about 20 ohm-cm. Thus the membrane is a poor conductor.

If the second statement is correct, the resting potential of the axon should approximately obey the Nernst equation for the potassium equilibrium potential, $V_{e,K} = (kT/e)\ln(n_{o,K}/n_{i,K})$. In Table 1.1 the composition of the axoplasm and blood of squid are given; using the values for potassium concentrations, the Nernst equation predicts a potassium equilibrium potential of -75 mV. Moore and Cole (1960) report squid axon resting potentials on the order of -70 mV; the difference between the two values may be attributed to the existence of a small permeability of the membrane to other ions.

A further consequence of 2 above is that the membrane should act like a potassium electrode with respect to variation in either $n_{o,K}$ or $n_{i,K}$. This has been confirmed by Curtis and Cole (1942) and Hodgkin and Keynes (1955) for $n_{o,K} > 20$ mM/L. The fact that at lower concentrations the membrane does not behave like a potassium electrode is a further indication of the non-zero permeability of the resting membrane to ions other than potassium. Indeed, use of the Goldman constant field equation (see Chapter 7; Goldman, 1943; Hodgkin and Katz, 1949) with $P_K:P_{Na}:P_{Cl} = 1:.04:.45$ gives a much better fit to the observed variations (P_i denotes the permeability of the membrane to an ion of type i).

In perfused axons Baker, Hodgkin, and Shaw (1961) found that the membrane behaves like a potassium electrode at low $n_{i,K}$. Deviations occur at concentrations above about 100 mM/L. These results may be explained by variations in the potassium permeability with membrane potential as well as the increased role of sodium in determining the membrane potential.

If statement 3 above is correct the maximum reversal of the membrane potential during excitation should be no larger than the membrane equilibrium potential for sodium. Further, the maximum overshoot potential during an action potential should behave like the potential of a sodium electrode, given by the Nernst relation for sodium: $V_{e,Na} = (kT/e)\ln(n_{o,Na}/n_{i,Na})$. Using sodium concentrations given in Table 1.1 the overshoot of the action potential should not exceed +55 mV. Indeed in the intact axon the overshoot is not larger than this, normally being some 40 mV. Also, Hodgkin and Katz (1949) have shown that the potential at the peak of the action potential varies approximately with $n_{o,Na}$ as predicted by $V_{e,Na}$. Variation in $n_{i,Na}$ in the fluid perfusing squid axons (Baker, Hodgkin and Shaw, 1961) shows that the overshoot responds in the predicted manner.

Statement 3 also requires that the increased Na^+ permeability (P_{Na}) of the membrane during excitation be transient; abundant proof of this was given by the voltage clamp experiments of Hodgkin and Huxley (1952a,b,c,d) discussed in Chapter 4.

If, as the fourth statement suggests, propagation is accomplished by local current flow, then changes in the external resistivity (ρ_o) of the medium around the axon should have dramatic effects on the conduction velocity of the impulse. Hodgkin (1939) has shown this to be the case. Immersion of the axon in oil, thereby increasing the external resistance, decreases the conduction velocity. Conversely, short circuiting the external and/or internal medium over a length of axon gives a dramatic increase in conduction velocity.

The third and fourth statements imply that, as a result of the large increase in the membrane Na^+ permeability and corresponding current flow, there should be an increase in the conductance of the membrane during activity. This has been adequately demonstrated by many investigators, but was first shown for the large excitable plant cell Nitella, and also squid, by Cole and Curtis (1938, 1939).

B. Strength-Duration Relation, Accomodation. Suppose we have the situation depicted in Figure 2.1 and the cell is a squid giant axon with a voltage recording microelectrode inserted at the site of current injection ($x = 0$). If finite duration constant current pulses are applied, the membrane potential record illustrated in Figure 3.1 will be obtained. In a similar fashion for inward (positive or depolarizing) currents below a certain value, cable-like responses are again obtained. However, once the current surpasses a given value, a complex voltage response (the action potential) occurs that reaches some +40 mV at its peak, and which propagates uniformly away from the site of current injection.

If the above experiment is repeated with current pulses of different durations, it is found that the amount of current necessary to elicit an action potential is a variable and dependent on the duration. Indeed, careful experiments such as described above yield a relation between the current (I) effective in eliciting an action

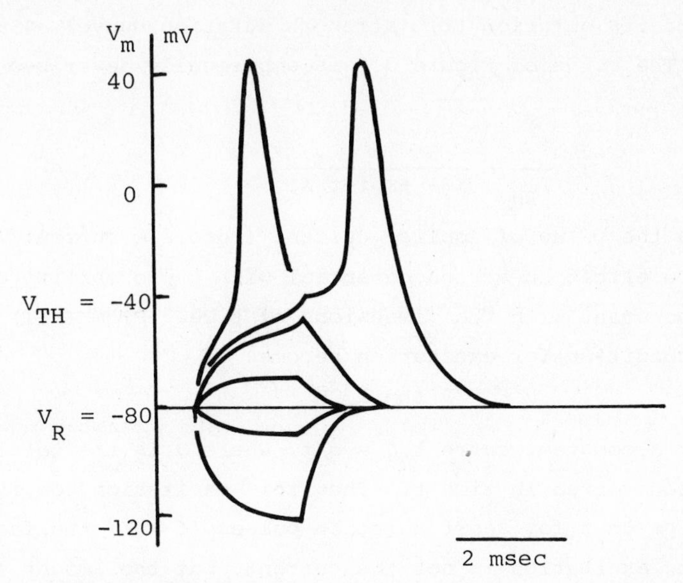

Figure 3.1 Membrane potential responses to a 2 millisecond duration current pulse. Hyperpolarizing currents and small depolarizing currents give normal cable responses. Larger depolarizing currents that lead to V_m exceeding threshold (V_{TH}) initiate action potentials. The time to initiation decreases as the magnitude of the applied current increases.

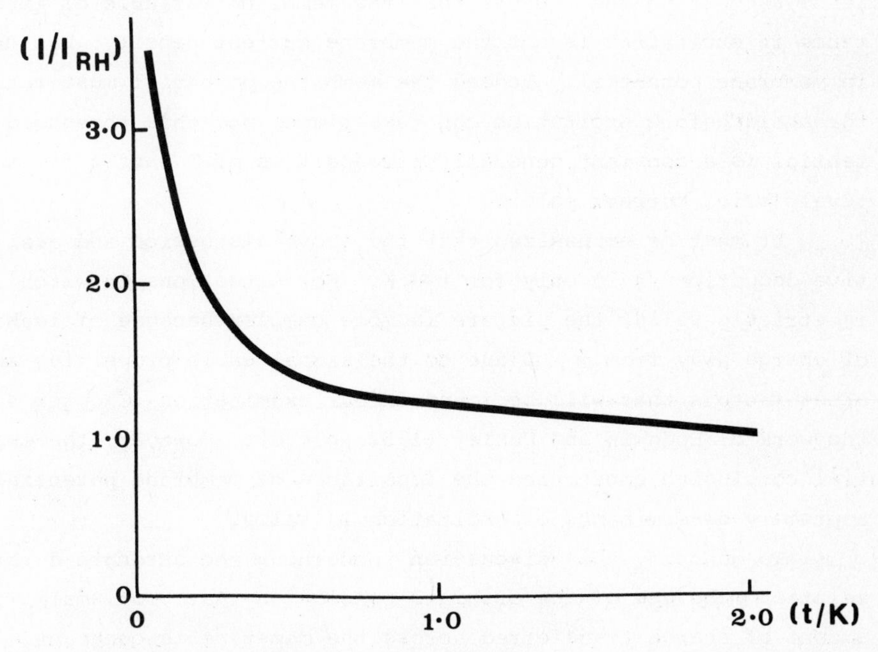

Figure 3.2 Strength-duration curve for an axon, according to equation 3.1. A long current pulse of magnitude I is applied, eliciting an action potential in time t.

potential and its duration (t) (strength duration curve) as shown in Figure 3.2. The curve of Figure 3.2 is empirically described by the equation

$$\frac{I}{I_{RH}} = \frac{1}{1 - \exp(-t/K)} \tag{3.1}$$

where I_{RH} is the value of applied current (rheobase current) just sufficient to elicit an action potential with a probability of 0.5 and K is a constant with the dimensions of time. From (3.1), when $t \ll K$, the condition for excitation becomes

$$I\ t = I_{RH}K$$

However, for a constant current, $J = q/t$, where Q is the total amount of charge transferred in time t. Thus this excitation condition really implies that for short duration pulses ($t \ll K$) the factor important for excitation is not the current, but the amount of charge transferred across the membrane. From previous discussion about the membrane electrical equivalent circuit the first effect of a transfer of charge across the membrane is to change the potential across the membrane capacity by a certain amount. This observation lends support to the concept that the membrane variable of importance in excitation is not the membrane current density, but changes in membrane potential. Indeed the membrane potential must reach threshold before excitation can take place, and this threshold potential is a constant generally irrespective of I and t for relatively brief current pulses.

It must be emphasized that the above discussion and qualitative deductions hold only for $t \ll K$. For situations in which (3.1) is strictly valid, the picture is more complex because of leakage of charge away from x = 0 due to the axonal cable properties and other factors that will be found in our examination (Chapter 4) of the work of Hodgkin and Huxley (1952a,b,c,d). However, the essential conclusion concerning the importance of membrane potential as a primary determinant of excitation is valid.

To quantify this discussion concerning the strength-duration relation make use of the original conclusion that, for small T, the amount of charge transferred across the membrane is constant. Thus, I want to compute the total charge transfer Q(T) due to I as a function of time. Clearly for an infinite cable

$$Q(T) = 2c_m \int_0^\infty \bar{V}_m(x,t)\,dx = 2\lambda c_m \int_0^\infty \bar{V}(X,T)\,dX \tag{3.2}$$

and thus, from Chapter 2,

$$\begin{aligned} \mathcal{L}\left[Q(T)\exp T\right] &= 2\lambda c_m \int_0^\infty u(X,s)\,dx \\ &= \lambda^2 I\, c_m(r_o+r_i)\left[\frac{1}{s-1} - \frac{1}{s}\right] \end{aligned} \tag{3.3}$$

so

$$\begin{aligned} \Delta Q(T) &= \lambda^2 I\, c_m(r_o+r_i)\left[1-\exp(-T)\right] \\ &= r_m c_m I \left[1-\exp(-T)\right] \end{aligned} \tag{3.4}$$

with $I_{RH} = \Delta Q(T)\tau_m^{-1}$ gives the strength duration relation directly. If instead of passing rectangular current pulses through the membrane, applied current that is a linearly increasing function of time leads to a different situation. As the rate of rise of current is decreased the magnitude of current required to initiate an action potential and the time required for initiation are increased. Eventually a situation would be reached in which (dI/dt) was so small that excitation would never occur. This process is called accommodation because the membrane appears to accommodate to the stimulating current in such a way that excitation is either delayed or never takes place.

Again, some insight into the fundamental processes leading to accommodation may be obtained by considering the cable properties of cells. A necessary precursor to excitation is the lowering of the membrane potential (depolarization) sufficiently to reach threshold. For linearly increasing currents, the larger (dI/dt) the faster the charge on the membrane capacity (and thus the membrane potential) will change. In concert with this change in charge on the membrane capacity is the tendency of the axon to restore its original charge distribution through the action of its cable properties. If a current with a sufficiently small (dI/dt) is employed the axon is able to resist excitation through its cable properties and through other factors discussed by Hodgkin and Huxley, predominantly inactivation of the membrane Na conductance.

C. Refractory States and Firing Frequency. Another property of excitable cells is the refractory period. If a suprathreshold conditioning current pulse is applied to an axon, thereby eliciting an action potential, and followed by a test pulse t milliseconds later two features will be noted. The first is that there exists a period of time following the generation of the first action potential during which the test pulse is completely unable to initiate a second action potential no matter how large it is. This period of time is the absolute refractory period often denoted by t_a. The second observation is that following the absolute refractory period there is a period in which the test pulse will initiate a second action potential but with the axon exhibiting an increased threshold that gradually returns to normal. This period of apparent increased threshold is called the relative refractory period.

It is easy to see that if the absolute refractory period of a fibre is t_a and the strength duration relation is given by

$$I = \frac{I_{RH}}{1 - \exp(-t/K)}$$

for a uniformly polarized axon then the time between action potentials with a constant current stimulus I will be

$$t - t_a = K \ln \frac{I}{I - I_{RH}} \qquad (3.5)$$

to predict as a first approximation that the impulse frequency f (impulses/sec) should be given by

$$f = \frac{1}{t} = \frac{1}{t_a + K \ln \dfrac{I}{I - I_{RH}}} \qquad (3.6)$$

If again, as before, K is arbitrarily taken to be τ, $T = t/\tau$, $T_a = t_a/\tau$, and $F = \tau f$ then

$$F = \left[T_a - \ln\left(1 - \frac{I_{RH}}{I}\right)\right]^{-1} \qquad (3.7)$$

It is clear from either (3.6) or (3.7) that the cell would be expected to have zero output (f or F = 0) for $I \leq I_{RH}$, and for $I > I_{RH}$ the firing frequency monotonically increases to an asymptotic value of t_a^{-1}.

Now

$$\ln(1 + Z) = Z - \frac{Z^2}{2} + \frac{Z^3}{3} - \ldots\ldots \qquad |Z| \leq 1$$

so from (3.7)

$$F = \left[T_a + \frac{I_{RH}}{I} + \frac{1}{2}\left(\frac{I_{RH}}{I}\right)^2 + \ldots. \right]^{-1}$$

$$\simeq \frac{I}{T_a I + I_{RH}} \qquad (3.8)$$

to predict that the firing frequency versus input relation should have the form of a rectangular hyperbola for $I \gg I_{RH}$.

D. Axonal Characteristics and Action Potential Propagation. Velocity. Action potentials, in contrast to subthreshold voltage responses, propagate without any decrement in their maximum amplitude. The action potential shape is variable about the stimulation site, but this variability vanishes one space constant away from the stimulus site. Thus for $x \gg \lambda$ the action potential is a travelling wave and satisfies the wave equation $(\partial^2 V_m/\partial x^2) = \theta^{-2}(\partial^2 V_m/\partial t^2)$ where θ is the propagation velocity. θ is about 20 m/sec for the non-myelinated squid giant axon.

To examine the effects of fibre parameters on the non-myelinated wave velocity predicted from axonal cable properties, suppose that instead of passing a constant current step into the axon as before the charge applied to the membrane is given by

$$Q = \int_{-\infty}^{\infty} I\,\delta(t)\,dt$$

The only place that this enters into the solution of the cable equation is through equation 2.22 specifying $(\partial V_m/\partial x)$ at $x = 0 \pm$, and its subsequent use in determining the constant A. Thus if the applied current is given by $I\delta(t)$

$$I\,\delta(t) = 2\int_0^{\infty} j_m(x,t)\,dt$$

In terms of X, T, and U this may be rewritten

$$Ie^T\,\delta(T) = \frac{2}{\lambda(r_o+r_i)}\int_0^{\infty} \frac{\partial^2 U(X,T)}{\partial X^2}\,dX$$

and taking the Laplace transform yields

$$I = \frac{2}{\lambda(r_o+r_i)}\int_0^{\infty} \frac{\partial^2 u(X,s)}{\partial X^2}\,dX$$

or

$$A = \frac{I\lambda(r_o+r_i)}{2\sqrt{s}}$$

Thus the full Laplace transform of U(X,T) is

$$u(X,s) = \frac{\lambda I(r_o+r_i)}{2} \cdot \frac{\exp(-\sqrt{s}\,X)}{\sqrt{s}}$$

Taking the inverse transform and writing the expression for $V_m(X,T)$,

$$V_m(X,T) = \frac{\lambda(r_o+r_i)I}{2\sqrt{\pi T}} \exp\left[-\left(\frac{X^2}{4T} + T\right)\right] \tag{3.9}$$

results. (Graphs of $V_m(X,T)$ are shown in Figure 3.3 for X =1 ,2 and 3). How "fast" does this response propogate and how rapidly is it attenuated?

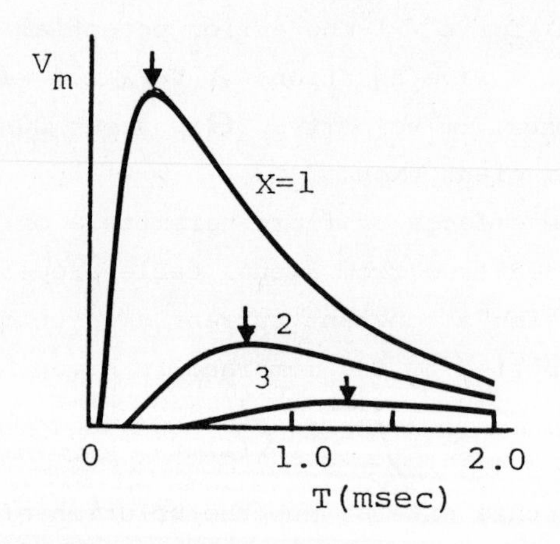

Figure 3.3 $V_m(X,T)$ versus T in response to a delta function current according to equation 3.9 at X = 1,2,and 3.

Denote the time at which $(\partial V_m/\partial T) = 0$ by $\bar{T}$. $\bar{T}$ will also depend on the distance from the site of current injection, so $\bar{X}$ is the location on the axon at which $(\partial V_m/\partial T) = 0$ at $T = \bar{T}$. From (3.3)

$$\frac{\partial V_m}{\partial T} = \frac{\lambda I(r_o+r_i)}{2\sqrt{\pi T}} \left[\frac{X^2}{4T^2} - \frac{1}{2T} - 1\right] \exp\left[-\left(\frac{X^2}{4T} + T\right)\right]$$

so by letting $(\partial V_m/\partial T) = 0$

$$\bar{X}^2 = 4\bar{T} + 2\bar{T} \tag{3.10}$$

the desired relation between $\bar{X}$ and $\bar{T}$ other than for $\bar{X}\to\infty$, $\bar{T}\to\infty$.

From equation 3.10 it is clear that when $\bar{T} \ll 1$, $\bar{X}^2 \simeq 2\bar{T}$ so the pulse (3.9) has a dimensionless velocity $V \propto 1/\sqrt{\bar{T}}$. When $\bar{T} \gg 1$,

$\bar{X}^2 \simeq 4\bar{T}^2$ and the velocity approaches a constant value given by $V = (\bar{X}/\bar{T}) = 2$. In terms of membrane related parameters the velocity is given by

$$\Theta = \frac{\bar{X}}{\bar{T}} = 2\lambda/\tau_m \qquad \text{(cm/sec)} \qquad (3.11)$$

For many invertebrate non-myelinated axons, $\tau_m \simeq 1$ msec and $\lambda \simeq 5$ mm to predict a velocity of 10^3 cm/sec or 10 m/sec for subthreshold responses.

In terms of specific fibre parameters the velocity becomes

$$\Theta = \tau_m^{-1}\sqrt{\frac{DR_m}{\rho_o+\rho_i}} \qquad (3.12)$$

where $D = 2a$ is the axon diameter. This relation predicts that the velocity of propagation in an unmyelinated fibre should vary as $\sqrt{D}$, as found experimentally.

To examine the role of myelinated fibre cable properties on action potential I must deal with (2.38). For simplicity, transform the dependent and independent variables as follow:

$$T = \alpha t/\tau$$

where

$$\tau = \left(\frac{r_m r_s}{r_m+r_s}\right)(c_m+c_s),$$

$$\alpha = \frac{r_m r_s}{(r_o+r_i)(r_m+r_s)},$$

and

$$U(n,T) = V_m(n,T)\exp T$$

so (2.38) may be rewritten as

$$U(n+2,T) - 2U(n+1,T) + U(n,T) = \frac{\partial U(n+1,T)}{\partial T} \qquad (3.13)$$

If the Laplace transform of $U(n,T)$ is denoted by $u(n,s)$ then (3.13) yields

$$u(n+2,s) - (2+s)u(n+1,s) + u(n,s) = 0 \qquad (3.14)$$

Assuming a solution of the form

$$u(n,s) = A(s)r^n$$

then it is clear that r must satisfy the equation

$$r^2 - (2+s)r + 1 = 0$$

Thus the two roots, r_1 and r_2, are given by

$$r_1, r_2 = \frac{(2+s) \pm \sqrt{(2+s)^2 - 4}}{2}$$

Note that $r_1 r_2 = 1$ and $r_1 + r_2 = 2 + s$ so I define a new quantity Γ by $r_1 = \exp(-\Gamma)$ and thus $r_2 = \exp\Gamma$.

From the above, the full solution of (3.14) will be given by

$$u(n,s) = A_1(s)\exp(-n\Gamma) + A_2\exp(n\Gamma) \tag{3.15}$$

and since it is required that $u(n,s)$ remain finite for n large, $A_2(s) = 0$. It remains now to determine the value of $A(s) = A_1(s)$.

The determination of $A(s)$ proceeds as before. At $n = 0$ a current $I\delta(t)$ is injected so

$$I\delta(t) = \sum_{-\infty}^{\infty} \left[j_m(n,t) + j_s(n,t)\right]$$

$$= 2\sum_{0}^{\infty} \left[j_m(n,t) + j_s(n,t)\right] - j_m(o,t) \tag{3.16}$$

However, from (2.31)

$$j_m(n,t) + j_s(n,t) = \frac{1}{r_o+r_i}\left[V_m(n+1,t) - 2V_m(n,t) + V_m(n-1,t)\right]$$

or in terms of $U(n,t)$

$$\left[j_m(n,t)+j_s(n,t)\right]e^T = (r_o+r_i)^{-1}\left[U(n+1,T)-2U(n,T)+U(n-1,T)\right] \tag{3.17}$$

Combining (3.16) and (3.17) yields

$$I\delta(T)e^T = 2(r_o+r_i)^{-1}\sum_{0}^{\infty}\left[U(n+1,T)-2U(n,T)+U(n-1,T)\right] -$$

$$(r_o+r_i)^{-1}\, U(1,T)-2U(0,T)+U(-1,T) \tag{3.18}$$

Take the Laplace transform of (3.18) and use (3.15) with $A_2(s) = 0$ and $A(s) = A_1(s)$ to obtain

$$I = A(s)(r_o+r_i)^{-1}\left\{2\sum_{0}^{\infty}\left[e^{-(n+1)\Gamma} - 2e^{-n\Gamma} + e^{-(n-1)\Gamma}\right] - \left[e^{-\Gamma} - 2 + e^{\Gamma}\right]\right\}$$

$$= -2A(s)(r_o+r_i)^{-1}\sqrt{s^2+4s}$$

so

$$A(s) = -\frac{I(r_o + r_i)}{2\sqrt{s^2 + 4s}}$$

Thus I may immediately write

$$u(n,s) = -\frac{I(r_o+r_i)}{2\sqrt{(2+s)^2-4}}\left[\frac{(2+s)-\sqrt{(2+s)^2-4}}{2}\right]^n$$

where I have used r_1 in place of $\exp(-\Gamma)$. To obtain U(n,t), note that

$$U(n,T) = \mathcal{L}^{-1}\left[u(n,s)\right] = e^{-2T}\mathcal{L}^{-1}\left[u(n,s-2)\right]$$

where

$$u(n,s-2) = -\frac{I(r_o+r_i)}{2\sqrt{s^2-4}}\left[\frac{s-\sqrt{s^2-4}}{2}\right]^n$$

Further (Roberts and Kaufman, 1966),

$$\mathcal{L}^{-1}\left\{\frac{\left[s-\sqrt{s^2-a^2}\right]^m}{a^m\sqrt{s^2-a^2}}\right\} = I_m(aT)$$

where $I_m(aT)$ is a modified Bessel function defined by

$$I_m(aT) = \sum_{n=0}^{\infty}\frac{(aT/2)^{m+2n}}{n!\,\Gamma(m+n+1)}$$

and $\Gamma(m+n+1)$ is the Gamma function defined by

$$\Gamma(x) = \int_0^{\infty} e^{-u}u^{x-1}du$$

With all of the above,

$$U(n,T) = -\frac{I(r_o+r_i)}{2}e^{-2T}I_n(2T)$$

or, in terms of $V_m(n,T)$

$$V_m(n,T) = -\frac{I(r_o+r_i)}{2}e^{-3T}I_n(2T)$$

To obtain estimates of the propogation velocity of this disturbance note that $I_n(2T)$ has the following properties:

$$I_n(2T) \simeq \begin{cases} \dfrac{T^n}{n!} & T \ll 1 \\ \dfrac{e^{2T}}{2\sqrt{\pi T}}\left(1-\dfrac{4n^2-1}{16T}\right) & T \gg 1 \end{cases}$$

Thus

$$V_m(n,T) \simeq \begin{cases} \dfrac{e^{-3T}T^n}{n!} & T \ll 1 \\ \dfrac{e^{-T}}{2\sqrt{\pi T}}\left(1-\dfrac{4n^2-1}{16T}\right) & T \gg 1 \end{cases}$$

What are the values $\bar{n}$ and $\bar{T}$ such that $(\partial V_m/\partial T)_{\bar{n},\bar{T}} = 0$? i.e. where is the peak of the propogating response? For small T, $\bar{n}$ and $\bar{T}$ are related by $\bar{n} = 3\bar{T}$ or in terms of normal variables

$$\bar{n} = 3\alpha\bar{t}/\tau$$

to give a propogation velocity θ (internodes/sec) of

$$\theta = 3(r_o + r_i)^{-1}(c_m + c_s)^{-1}$$

If the axon diameter at the node is d cm and the node width is w cm then the nodal area is π dw and

$$c_m = \pi dwC_m$$

Further if the internode distance is L cm,

$$r_i = 4L\rho_i/\pi D^2$$

and

$$c_s = \pi DLC_s$$

to give, with $r_o \ll r_i$,

$$\theta = \frac{3D^2}{4L\rho_i(dwC_m + DLC_s)}$$

There is abundant experimental evidence (c.f. Rushton, 1951) that the function $g(D) = d/D$ is a slightly increasing function of D with $g(D) \simeq 0.6$ on the average. Further, experimentally $f(D) = (L/D) \propto g(D)[-\ln g(D)]^{1/2}$ or $f(D) \simeq K$ where K is a constant. Thus the conduction velocity $\bar{\theta} = \theta L$ (cm/sec) may be written as

$$\bar{\theta} = \frac{3D}{4\rho_i(0.6wC_m + KDC_s)} \tag{3.19}$$

to predict a linear relation between θ and D as found experimentally For large T the relation $\bar{n}^2 \simeq 8\bar{T}^2/3$ holds so

$$\bar{\theta} = \frac{D\sqrt{8/3}}{4\rho_i(0.6wC_m + KDC_s)} \tag{3.20}$$

PROBLEMS

3.1. Verify that the solutions to the problem of determining the membrane potential response to an applied delta function current are as given in the text for myelinated and non-myelinated fibres.

3.2. Write $V_m(X,T)$ and $V_m(n,T)$ in terms of $V_m(0,T)$ from (3.9) and (3.18a) respectively. Plot $V_m(X,T)/V_m(0,T)$ and $V_m(n,T)/V_m(0,T)$ for $X = 1,2,3$ and $n = 1,2,3$ respectively.

3.3. Compare and contrast the role of myelinated and nonmyelinated fibre structure in determining conduction velocity, threshold effects, the strength duration relation, and axonal firing frequency in response to an external current input.

CHAPTER 4. THE HODGKIN-HUXLEY AXON

To this point I have touched only briefly on the ionic events that occur during the generation and propogation of an action potential. These events were first elucidated by Hodgkin, Huxley and Katz almost three decades ago through a powerful technique (voltage clamp) developed and perfected by K.S. Cole and G. Marmont.

In this chapter I discuss voltage clamping and the rationale behind its development, the data that have been obtained from it relating to membrane ionic currents, the Hodgkin-Huxley empirical description of these currents, and finally their reconstruction of the action potential based on knowledge of the time and voltage dependent behavior of the membrane conductances. More extensive discussion of these topics is to be found in the excellent book by Hodgkin (1964).

A. Voltage Clamping. In Chapter 3, I characterized the action potential as an all-or-none process that continued to completion once the membrane potential has passed a threshold level. Since it was argued from experimental data that the apparent critical state variable is the membrane potential, and not membrane current, it is clear that a technique for placing the membrane potential under experimental control is essential for elucidating the processes underlying the generation of the action potential. A second problem is that the action potential does not stand still for the experimenter, but rather appears as a traveling wave. Without a doubt, both the propagating and all-or-nothing characteristics were foremost in the thoughts of Cole (1949) and Marmont (1949) when the voltage clamp technique was developed.

In its simplest form, voltage clamping is nothing more than the process of placing an ideal (zero internal impedance) variable potential battery across the nerve membrane. The experimenter sets the potential of the battery (actually a sophisticated electronic feedback system), and the battery supplies the current through the membrane necessary to maintain the desired potential.

To tabulate the currents generally flowing through the membrane, it is necessary to extend the treatment of Chapter 2 slightly. The membrane current, j_m, will be composed of two components - the displacement current, $c_m(\partial V_m/\partial t)$ as before, and an ionic current $j_i=j_i(V_m,t)$ to be experimentally determined (not assumed to have the ohmic behavior of Chapter 2). This membrane current is identically equal to the sum of the currents supplied from adjacent regions of the nerve fibre, $(r_i+r_o)^{-1}(\partial^2 V_m/\partial x^2)$ and any current supplied from any external source, j_c, such as the voltage clamp. Thus

$$j_c + \frac{1}{r_i+r_o}\frac{\partial^2 V_m}{\partial x^2} = c_m\frac{\partial V_m}{\partial t} + j_i(V_m,t) \qquad (4.1)$$

Equation 4.1 makes the principle of voltage clamping transparent. It is first experimentally arranged that r_o is small (by, for example, placing the fibre in a large bath) and then inserting a wire inside the axon along its axis. The result of this procedure is to eliminate spatial variations in potential, effectively 'space clamping' the membrane by placing all of the membrane in direct electrical contact, with the resultant identical appearance of any membrane potential variation at all points on the axon. If the fibre is stimulated at this juncture, a 'membrane action potential' is produced.

Once the fibre is space clamped, (4.1) becomes

$$j_c = c_m\frac{\partial V_m}{\partial t} + j_i(V_m,t) \qquad (4.2)$$

and thus the applied current is identical to the membrane current. Use now of the ideal battery (voltage clamp), capable of producing any $V_m(t)$, will allow the immediate calculation of j_i simply by recording j_c and calculating $\dot{V}_m(t)$. Things become particularly simple if it is arranged that V_m is independent of time, for in this case the applied current is identical to the ionic current. This is the usual procedure, although other variations of $V_m(t)$ have been used.

This then is the theoretical basis of voltage clamping. The practical implementation of the procedure is, however, technically quite difficult. I will deal with none of these technical complications here, and the interested reader should consult Cole and Moore (1960), Moore and Cole (1963), and Taylor et al. (1960) for a detailed discussion.

B. Ionic Current Flow as Revealed by the Voltage Clamp. Drs. A.L. Hodgkin and A.F. Huxley married the voltage clamp technique with the squid giant axon in successful attempts to determine the underlying ionic basis of the action potential. Their work (Hodgkin and Huxley, 1952a,b,c,d) was recognized for its far-reaching implications by the Nobel Prize Committee in 1964.

As I pointed out above, the total current flowing through the axon membrane in response to a step depolarization of hyperpolarization is composed of two parts, a surge of capacity current and an ionic component. Under voltage clamp conditions, the displacement current is over in a very short period of time and what one sees is mainly ionic current. With voltage clamp control, records of ionic current versus time similar to those shown in Figure 4.1a are obtained. The general characteristics of these current records can be summarized by saying that there is generally an initial inward flow of current followed by a later outward flow of current. For some ranges of membrane potentials, there is no indication of an inward current flow (i.e. the currents are entirely outward) while for other ranges of clamping potentials the currents are entirely outward.

Hodgkin and Katz (1949) speculated that qualitatively what happened during an action potential was an initial transient increase in P_{Na} followed, perhaps, by an increase in P_K. They thought that P_{Na} was increased by a membrane depolarization, which in turn leads to an influx of Na^+ and further depolarization, and thus further increases in P_{Na}. The exact nature of these changes, e.g., their possible dependence on membrane potential and time, was unclear. Therefore, it seemed possible to Hodgkin and Huxley that the early inward current noted in some records might also be due to sodium. To test this hypothesis two experimental techniques were employed. In the first, the external sodium concentration was reduced to zero by replacement with choline, and it was found that the early inward current disappeared at all membrane clamping potentials. The second check on the hypothesis that the early currents are due to sodium was based on the fact that at the sodium equilibrium potential there should be no flow of sodium ions across the membrane. When the membrane potential was clamped at $V_{e,Na}$, as estimated from the Nernst equation, it was found that the early current did indeed disappear.

Hodgkin and Huxley made the assumption that the ionic current was made up of the sum of sodium and potassium ion currents and that there was no interaction between them, and so were able to determine the component currents j_K and j_{Na}. In Figure 4.1b I illustrate the procedure whereby they separated the total ionic current into its components. Recorded current in normal sea water is assumed to be $j_K + j_{Na}$, and the current recorded in choline Ringers solution is assumed to be j_K alone. Thus the sodium current is given as the difference between the currents in normal sea water and the current carried in sodium deficient sea water. In Figure 4.1b I have indicated what $j_K(t)$ and $j_{Na}(t)$ at one depolarizing clamping potential would look like using this procedure. This process may be carried out at a number of clamping potentials to give a family of current versus time records.

Rather than dealing with the specific ionic currents directly, Hodgkin and Huxley chose to take the chord conductance $g_c=j/(V_m-V_e)$ as a measure of the ease with which either sodium or potassium traversed the membrane in response to a given membrane potential and combination of concentration gradients. Thus, they defined chord conductances for sodium and potassium by $g_{Na}=j_{Na}/(V_m-V_{e,Na})$ and $g_K=j_K/(V_m-V_{e,K})$ respectively. They were able to use their separated sodium and potassium current records to derive equivalent chord conductance as a function of time for various membrane potentials as illustrated in Figure 4.1b for one depolarization.

In that figure note that the potassium conductance rises in a sigmoidal fashion and that there is a delay in its rise. The degree of delay decreases as the depolarization is increased. The potassium

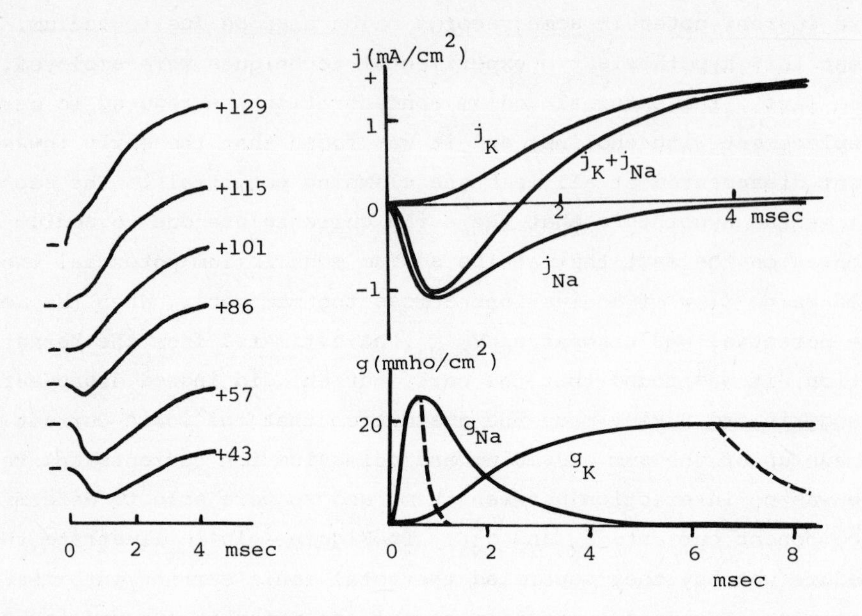

Figure 4.1 a. A series of six ionic current versus time records obtained by voltage clamping the squid giant axon membrane by the indicated depolarizations (in millivolts) from V_R. Outward currents are indicated as upward deflections. The 101 mV depolarization took V_m to $V_{e,Na}$. b. Upper. The total and seperated ionic currents for a 57 mV depolarization. The middle record is in normal sea water, the upper is in sea water with Na replaced by choline (assumed to be impermeable), and the bottom record is the difference between the two. Lower. Derived chord conductances for the sodium and potassium currents of the upper part of the figure (note the change in time scales). The dotted lines indicate the exponential decrease that would be found in the conductances if the 57 mV clamping pulse was turned off at the peak of g_{Na}, or at 6 msec.

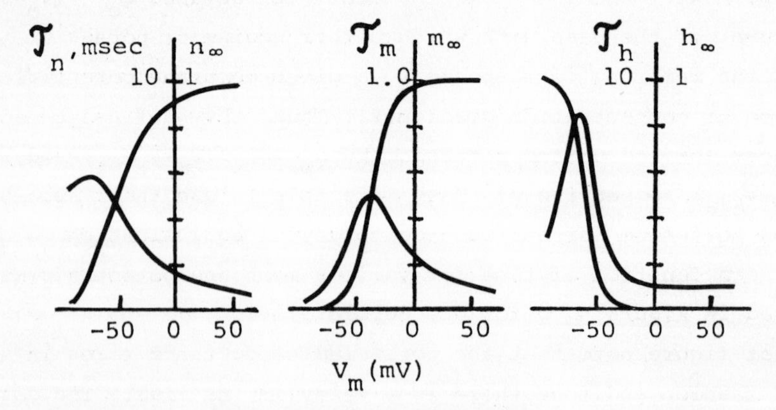

Figure 4.2 s_∞ and τ_s (s=m,n,h) versus V_m. See text.

conductance maintains a steady value for a period of at least ten milliseconds at all depolarization levels. The situation for the sodium conductance is quite different. The first point to be noted is that in response to a maintained depolarization there is little delay in the rise of the sodium conductance. Further, the sodium conductance does not stay at its maximal level but spontaneously declines to a level almost equal to that of its resting level. This phenomenon, termed inactivation of the sodium conductance, has two important features: a) The time at which inactivation begins is strongly dependent on the level of depolarization, decreasing for increasing depolarizations; and b) The rate of fall of the sodium conductance during its inactivation phase is increased as the level of depolarization is increased.

The characteristics of the potassium and sodium conductances demonstrated in Figure 4.1b do not tell all of the story. If, instead of maintaining a depolarization for a long period of time the membrane is hyperpolarized after a brief depolarization, both the potassium and sodium conductances instantaneously start to decay in an exponential fashion when the membrane is repolarized to its initial resting level. The response of the membrane, and the associated sodium and potassium chord conductances, upon repolarization is thus in sharp contrast to the characteristics noted when the membrane is initially depolarized.

The phenomenon of the inactivation of the sodium conductance, discussed above, is an extremely important one in terms of understanding how the normal nerve axon functions. The second characteristic of the inactivation of the sodium conductance that is not revealed by the experiments described above can be demonstrated with the following experimental situation. If two pulses are applied to the membrane in succession and the duration of the first pulse is varied, the conditioning (first) pulse will elicit a given change in the sodium conductance of the nerve. The second (test) pulse will reveal the time course of changes in the inactivation of the sodium conductance. Depolarizing conditioning pulses "inactivate" the sodium system while hyperpolarizing conditioning "activates" it. This activation and inactivation "accumulates" and decays exponentially back to its normal value.

These are the primary features of ionic current movements through normal axons as functions of both membrane potential and time. From these features, one can qualitatively hypothesize about the sequence of events underlying the generation and propagation of an action potential.

From the treatment (Chapter 2) of the subthreshold electrical properties of cells, a small depolarization at a point on the membrane will produce a depolarization of the surrounding membrane. The depolarization produced is a decreasing function of the distance away from the initial site of depolarization, and the rate of decrease is determined by the axonal space constant, λ. In the depolarized region of the membrane there will be an increase in g_{Na} and hence an inwardly directed j_{Na}, constituted in large part of charge drawn from surrounding areas of the membrane (thus creating an outward current in those regions). If the initial depolarization is such that the increase in g_{Na} makes it the predominant conductance in the membrane, as opposed to the resting state where $g_{Na} \ll g_K$, a type of regenerative (positive feedback) phenomena will occur. The increase in g_{Na} leads to an increase in j_{Na}, which further depolarizes the axon, leading to increased g_{Na}, etc. (This process, however, will continue only as long as $V_m < V_{Na}$). If this was the only process operating the membrane potential would be expected to approach V_{Na}, and then return to its resting level as g_{Na} inactivates.

However, there is a second factor to be taken into account in this qualitative reconstruction of the action potential, which is the delayed rise in g_K after g_{Na} starts to fall. As g_K increases, there will be an increased outward j_K until $|j_K| > |j_{Na}|$ and the net ionic current is outwardly directed. Now the outward j_K is opposing the inward j_{Na} and the fall in V_m due to the inactivation of g_{Na} is more rapid than expected. Eventually V_m will be determined solely by j_K and g_K, when g_{Na} is completely inactivated and j_{Na} is zero. At this point g_K will have been reduced to a value considerably below its maximum because of the repolarization (due principally to g_{Na} inactivation) and will be continuing to decrease. j_K will thus continue to decrease, the whole cycle being repeated until V_m has returned to its resting value.

C. Empirical Formulae for the Conductances. Hodgkin and Huxley (1952d) succinctly summarized their experimental findings on the voltage and time dependence of $g_{Na}(V_m,t)$ and $g_K(V_m,t)$ by a set of empirical formulae that have no obvious physical basis. They defined

$$g_{Na}(V_m,t) = \bar{g}_{Na}m^3h \tag{4.3}$$

and

$$g_K(V_m,t) = \bar{g}_K n^4 \tag{4.4}$$

wherein $\bar{g}_{Na} = 120$ mmho/cm^2 and $\bar{g}_K = 36$ mmho/cm^2 are constants and the variables m,n, and h are given as solutions of the equations

$$\frac{ds}{dt} = \alpha_s(V_m)(1-s) - \beta_s(V_m)s \qquad s = m,n,h \tag{4.5}$$

In the equations 4.5 the six 'rate constants' α_s, β_s were determined from data at 6.3°C as

$$\alpha_m = \frac{0.1(V_m+25)}{\exp\left(\frac{V_m+25}{10}\right)-1}, \qquad \beta_m = 4\exp(V_m/18)$$

$$\alpha_n = \frac{0.01(V_m+10)}{\exp\left(\frac{V_m+10}{10}\right)-1}, \qquad \beta_n = 0.125\exp(V_m/80)$$

$$\alpha_h = 0.7\exp(V_m/20), \qquad \beta_h = \frac{1}{\exp\left(\frac{V_m+30}{10}\right)+1}$$

I find it more instructive to write equations 4.5 in the form

$$\frac{ds}{dt} = -(s-s_\infty)/\tau_s \tag{4.6}$$

where $s_\infty = \alpha_s/(\alpha_s+\beta_s)$ is the steady state value of s at a particular membrane potential and $\tau_s = 1/(\alpha_s+\beta_s)$ is a 'time constant'. Utilizing the empirical expressions for α_s, β_s given above, I have calculated $s_\infty(V_m)$ and $\tau_s(V_m)$ for s = m,n, and h and the results are presented in Figure 4.2.

The effects of temperature on these empirically defined parameters are of interest. In biology it is customary to characterize the temperature dependence of a process M(T) by the Q_{10}, defined by $Q_{10,M} = M(T+10)/M(T)$ where temperatures are measured in $^\circ$C. Hodgkin Huxley, and Katz (1952) found that the temperature coefficient for the rate of rise of $j_{Na}(t)$ and $j_K(t)$ was on the order of 3, while the Q_{10} for the maximum in the currents at any given V_m was much smaller and about 1.1-1.3.

A number of heuristic schemes have been put forward as bases for the Hodgkin-Huxley empirical representation, and two may be found in Noble (1966) and Woodbury (1965). Agin (1963) has advanced an interesting and provocative derivation of the Hodgkin-Huxley equations based on quantum statistical considerations.

D. Reconstruction of the Action Potential. With the results summarized in the previous section Hodgkin and Huxley (1952d) were able to write the total ionic current as

$$j_i(V_m,t) = j_{Na} + j_K + j_l$$
$$= g_{Na}(V_m - V_{e,Na}) + g_K(V_m - V_{e,K}) + g_l(V_m - V_{e,l}) \qquad (4.7)$$

where j_l is an unspecified, but small, leakage current, and thus (4.1) should predict the form of $V_m(x,t)$ in response to an applied current j_c. Hodgkin and Huxley obtained numerical solutions to (4.1) for both space clamped situations and freely propagating action potentials and found their results closely matched experimental data. Further, their results accurately predicted such fibre properties as conduction velocity, refractory periods, time course of total membrane conductance change during an action potential, subthreshold oscillatory responses, accomodation, and others. Thus it must be concluded that they have adequately accounted for the major events underlying the generation and propagation of the action potential.

CHAPTER 5. CURRENT LEVELS OF KNOWLEDGE ABOUT THE EARLY AND LATE CURRENT FLOW PATHWAYS

The recent coupling of electrophysiological and pharmacological techniques in the study of excitation in the squid axon membrane and the node of Ranvier in frog myelinated nerve has yielded much information about the molecular mechanisms underlying the membrane changes during excitation.

A. The Action of Tetrodotoxin (TTX). TTX is a powerful neurotoxin derived from the liver, testes, and ovaries of the fugu (puffer) fish widely eaten in Japan. It is known that TTX has the molecular formula $C_{11}H_{17}N_3O_8$ and its structure is (Kao, 1966):

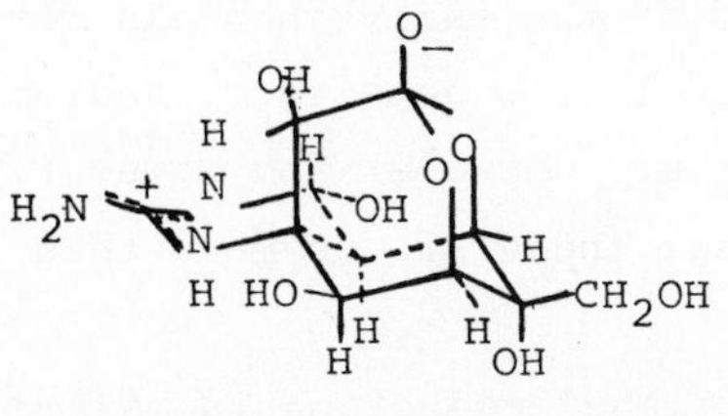

Figure 5.1 The structure of tetrodotoxin.

Little was known about the mode of action of TTX until Narahashi, Deguchi, Urakawa and Ohkubo (1960), and Nakajima, Iwasaki, and Obata (1962) surmised from the effects of TTX on lobster axon and frog muscle that the toxin acted principally to block excitation by a selective inhibition of the early sodium current. Further, in the presence of TTX Takata (1966) noted a progressive decrease in action potential height that was consistent with TTX interfering with early j_{Na}.

Voltage clamp experiments on single lobster and squid axons

(Narahashi, Moore, and Scott, 1964; Nakamura, Nakajima, and Grundfest, 1965; Takata, Moore, Kao, and Fuhrman, 1966; Hille, 1966; and Narahashi, Anderson, and Moore, 1967) verified the hypothesized action of TTX on the sodium system, and revealed much about the early current pathway in the process. Briefly, these investigators found that in nanomolar concentrations externally applied TTX reversibly reduces $g_{Na,\infty}$ $(=\bar{g}_{Na}n_{\infty}^{4})$ without affecting the time constants of the sodium activation and inactivation processes, τ_m and τ_n. TTX has no effect when it is in the perfusion fluid of internally perfused axons as shown by Narahashi, et al. (1967), nor does it affect the potassium and leakage pathways.

Current may be carried in either direction through the early current pathway in squid, depending on the experimental arrangement, and by a number of monovalent cations. Experiments by Watanabe, Tasaki, Singer, and Lerman (1967) and Moore, Blaustein, Anderson, and Narahashi (1967) have revealed that the action of TTX is independent of the direction of current flow through the sodium channel, or the type of cation (Li^+, Na^+, K^+, or Rb^+) carrying this current.

The quite specific action of TTX has yielded another interesting piece of information. Moore, Narahashi, and Shaw (1967) used TTX to determine the density of sodium channels in lobster nerve by a bio-assay technique under the assumption that one TTX molecule is capable of blocking one and only one of the early channels for current flow. They found a surprisingly low density of sodium channels, less than $13/\mu^2$ of membrane area. Chandler and Meves (1965) using voltage clamp data estimate that there are no more than 100 early current channels per μ^2.

Hille (1967) measured the fractional reduction in maximum j_{Na} as a function of TTX concentration, and concluded that there was a 1-1 binding between channel and TTX with an equilibrium constant of 1.2×10^{-9} M.

Villegas et al. (1971) isolated lipid components from the squid plasma membrane and demonstrated that the only component with which TTX interacts is cholesterol. They conclude that cholesterol is a component of the sodium channel that in some way determines g_{Na}.

Woodward (1964) has shown that three different forms of the

TTX molecule are possible - two are cationic and the third is a zwitterion. The relative proportions of these three fractions change as the pH of the medium containing TTX is varied, and Camougis, Takman and Tasse (1967) used this fact along with the known action of TTX on the nodes of the frog myelinated nerve fiber (Hille, 1966) to examine the relative potency of the three TTX forms. They found that the cationic forms of TTX were significantly more effective in blocking excitation than was the zwitterionic form. TTX contains a guanidinium group with a pK of about 11.5, but the changes giving rise to the cationic and zwitterionic forms also occur at the C4 and hemilactal oxygen points, indicating that these are important for the operation of TTX, as pointed out previously by Narahashi, Moore, and Poston (1967).

The data from studies using TTX indicate the following about the mechanisms of excitation:

1. TTX acts exclusively to block the flow of current through the early pathway in squid, irrespective of the direction of current flow or the cation carrying the ion, be it Li^+, Na^+, K^+ or Rb^+. Thus the conclusion that the early current channel is functionally distinct from the potassium and leakage channels seems likely.
2. TTX exerts its action on the early current flow by a simple decrease in the maximum conductance of that pathway. Since there is no alteration in the kinetic parameters of this pathway, it must be concluded that the molecular basis of these kinetics is either unaffected by TTX, or removed from activity. Further TTX does not significantly affect any molecular structure that may give rise to a potential to which the kinetic parameters are sensitive.
3. The relatively greater potency of the cationic forms of TTX indicates that at least part of its action is through interactions with a net negative charge on the outer surface of the membrane.
4. Cholesterol may be partially involved in the structure of the sodium channel.
5. The early current pathways are relatively sparse on the molecular level.

B. The Action of Tetraethylammonium Ion (TEA). Considerable interest in the action of tetraethylammonium ion on membrane properties existed even before the demonstration by Tasaki and Hagiwara (1957) that when TEA^+ was injected in the squid giant axon cardiac like action potentials were produced. Voltage clamp experiments by Armstrong and Binstock (1965) and Armstrong (1966) on squid axons indicate that internal (but not external) TEA^+ ions act primarily to prevent outward potassium current flow through the late current channel. TEA^+ does not block inward current flow through this channel, nor affect the kinetics of the early or late cuttent pathways or the leakage conductance. The use of TEA^+ permitted Armstrong (1966) to give an upper estimate of the number of potassium channels in squid, $69/\mu^2$. If the percent reduction in g_K is plotted as a function of TEA concentration, the "titration curve" obtained is consistent with a kinetic scheme wherein one TEA^+ is sufficient to block one potassium channel with a rate constant of 4×10^{-4} moles. Finally, studies on the effects of TEA^+ on the late current pathway in the presence and absence of TTX conclusively indicate no competition between TTX and TEA^+ for a common binding site. Thus, the conclusion that the early and late current flow pathways are distinct is given further support.

C. Ionic Strength Effects on the Kinetic Parameters of Excitation. Tasaki and Shimamura (1962), Baker, Hodgkin and Shaw (1962), Narahashi (1963) and Baker, Hodgkin and Meves (1964) all observed that if the squid axon is perfused with solutions diluted with isotonic sucrose, action potentials are obtained even though the resting potential is near zero. This result is not included in the Hodgkin-Huxley equations due to the high degree of sodium inactivation at low membrane potentials. As pointed out by Narahashi (1963) and Baker, Hodgkin and Meves (1964) the effect could, however, be explained by a shift of the sodium activation and inactivation versus membrane potential curves. Such shifts were found by Moore, Narahashi and Ulbricht (1964) and by Chandler, Hodgkin and Meves (1965), who further showed that there was a shift in the time constants of the sodium conductance activation and inactivation processes with respect to membrane potential. This translation along the voltage axis was correlated solely with changes in internal ionic strength.

Having found such shifts in the sodium parameters versus membrane potential curves, it would not be unexpected that changing internal ionic strength would have similar effects on the potassium system. Unfortunately, the experimental data are not clear cut for the potassium channel. Results presented by Moore, Narahashi and Ulbricht (1964) indicate that there may indeed be a shift in the potassium parameters with respect to membrane potential; but the data of Chandler, Hodgkin and Meves (1965) indicates that there may be anywhere from no shift in the potassium conductance characteristic to as much as 50% of the amount of shift seen in the characteristics of the sodium channel. Thus, the effects of low ionic strength solutions on the potassium channel are not clear.

A reasonable explanation of the effects of low ionic strength solutions was advanced by Baker, Hodgkin and Meves (1964) and quantified by Chandler, Hodgkin and Meves (1965). They speculate that negatively charged groups at the inner membrane surface affect the potential drop between the inner and outer surfaces of the membrane. The magnitude of the potential drop due to these charged groups will depend on the ionic strength of the medium bathing the membrane. They assumed that the sodium channel, in particular, is sensitive to the potential of the double layer at the inner border and the measured membrane potential. An adequate description of the observed effects for low ionic strength solutions can be made using the Verwey-Overbeck (1948) theory with a net excess negative charge density of 1.4×10^{13} charges/cm^2 on the inner surface of the membrane. This does not have to be uniform but must be over an area of sufficient extent surrounding each sodium channel.

The obvious conclusion from these studies is that the potential controlling the kinetic characteristics of the sodium channel, and possibly also of the potassium channel, is a combination of the trans-membrane potential of the axon and a potential produced by negative fixed charge in the membrane. Exception might be taken to the assumption of the existence of net negative charge in the membrane. However, the cation to anion permeability ratio of many cell membranes similar to that of the squid, coupled with electrophoretic studies on cell suspensions (see Woodbury _et al_., 1969 for references pertinent to these points) make the assumption highly plausible. If the fixed charge density is 1.4×10^{13} charges/cm^2, then these charges are separated by about 27 Angstroms in a rectangular array.

D. The Effect of Calcium on the Membrane. There are two especially useful experimental studies on the effects of calcium on the Hodgkin-Huxley parameters of large single axons. Frankenhaeuser and Hodgkin (1957) examined the effects of variations in $n_{o,Ca}$ in squid and Blaustein and Goldman (1966) repeated the same work on the lobster axon. The results of these two studies indicate that calcium exerts two distinct actions on the excitable parameters of the membrane:

a) an increase in external calcium concentration has the effect of shifting the sodium and potassium conductance versus potential curves in the depolarising direction. They have a similar effect on the kinetic parameters of the sodium activation system . The data for the effects of calcium on the potassium activation kinetics is meager, but indicates that qualitatively the effect is the same. Frankenhaeuser and Hodgkin found that a five-fold increase in external calcium concentration is equivalent to a 10-15 mV hyperpolarisation;

b) an increase in $n_{o,Ca}$ may decrease the maximum obtainable sodium and potassium conductances, $\bar{g}_{Na}$ and $\bar{g}_K$.

Huxley (1959), using the empirically observed effects of calcium variation, successfully modified the Hodgkin-Huxley equations to account for the oscillatory membrane potentials noted when external Ca^{++} is decreased.

The first effect of Ca^{++} listed above is so similar to the effects of internal ionic strength changes that it is impossible to resist proposing a similar mechanism for the two effects. Namely, changes in external calcium modify a local (zeta) potential at the membrane surface to which the kinetic parameters of excitation are sensitive. In order to produce the observed results, the potential would also have to be due to negative fixed charge. Calculations similar to those made by Chandler, Hodgkin and Meves (1965) indicate that the density of these negative fixed charges would have to be about $3.2 - 4.8 \times 10^{13}$ charges/cm^2. Based on the effect of ionic strength, Chandler estimated that the charge density is 1.4×10^{13} charges/cm^2. The two values are remarkably close. It would be attractive to propose that there is negative fixed charge at both the exterior and interior borders of the membrane.

Though the effects of external calcium on $\bar{g}_{Na}$ and $\bar{g}_K$ are not as definite as the shifts along the potential axis of these and other parameters, the results may be taken to indicate that calcium acts in a second and different manner. It does not seem unlikely that in addition to modifying the voltages to which membrane paramaters are sensitive, calcium can also interact with the molecular substructure responsible for determining $\bar{g}_{Na}$ and $\bar{g}_K$.

E. The Effects of pH Changes. The effects of altering solution pH are quite similar to effects produced by divalent ion concentration shifts. A reduction in pH has an effect qualitatively similar to that observed when $n_{o,Ca}$ is elevated (Stillman *et al.*, 1971). If one examines the reduction in $\bar{g}_K$ as pH is decreased, there is a sharp break at a pH = 4.5. This has been interpreted as resulting from the titration of a phosphoric group intimately associated with the potassium channel. Further, decreases in pH lead to a reduction in $\bar{g}_{Na}$ and the data are consistent with the involvement of a titratable group of pK = 6.5 in the determination of $\bar{g}_{Na}$. The likely guess is that this functional group is a carboxyl group.

F. Cation Selective Properties of the Membrane at Rest and During Excitation. Chemical modification of the squid axon membrane by tetrodotoxin and tetraethylammonium indicates that the early, late and leakage current pathways of the membrane are distinct and separate. This is consistent with the difference in electrical properties of the three current pathways. Studies on the ability of various cations to substitute for sodium and potassium during excitation tend to support these conclusions.

The cation permeability sequence for the early pathway for current flow during excitation, the sodium channel, is P_{Li}: P_{Na}: P_K: P_{Rb}: P_{Cs} = 1.1: 1.0: 0.083: 0.025: 0.016 (Chandler and Meves, 1964, 1965; Meves and Chandler, 1965). Senft and Adelman (1967) and Moore *et al.* (1966) found this same relative sequence for the sodium channel but with slightly different numbers. Tasaki, Singer and Watanabe (1966) found the relative permability sequence $P_{Na} > P_{NH4} > P_K > P_{Rb}$ for the sodium channel, and Binstock and Lecar (1969) have confirmed the relative position of the sodium and ammonium ions. Taking these results together a relative monovalent cation permeability sequence of $P_{Li} > P_{Na} > P_K > P_{Rb} > P_{Cs}$ holds for the sodium channel. The kinetic behaviour of currents carried by any of these ions through the sodium channel appear to be identical.

The potassium channel shows quite different cation selectivity properties from those of the sodium channel. Chandler and Meves (1965) and Meves and Chandler (1965) have shown that a number of other monovalent cations can pass through this region although less easily than potassium. The permeability sequence is $P_K : P_{Na} : P_{Rb} : P_{Cs} = 1:0.38:0.27:0.015$. Tasaki, Singer and Watanabe (1965, 1966) found the same sequence for K, Rb, and Cs, while Moore _et al_. (1966) found that rubidium is less permeable in the potassium channel than is potassium. Further, Binstock and Lecar (1969) showed that with respect to potassium the ammonium ion is one half as permeable in the potassium channel. The composite monovalent cation selectivity sequence for the potassium, or late current flow, channel in squid is $P_K > P_{NH_4} > P_{Na} > P_{Rb} > P_{Cs}$. As in the sodium channel, the kinetic machinery of the potassium channel appears to operate unchanged no matter what ion substitutes for potassium.

A decade ago there were few facts concerning the molecular basis of membrane discrimination between various cations. Research of recent years has yielded a wealth of data which any molecular theory for this discrimination must include. As explored in more depth elsewhere (Woodbury et al., 1969; Mackey, 1971) there is enough presumptive evidence from biochemical systems to imply that the selectivity of the sodium channel is controlled by an ionized carboxyl group, while the selectivity of the potassium channel is due to ionic interactions with a phosphoric group. Simple quantitative calculations utilizing molecular parameters confirm that this is physically consistent and reasonable.

G. Kinetic Behaviour in the K Channel. The dynamic properties of the potassium conductance system were explored in more detail by Cole and Moore (1960) who found that, although fourth order kinetics were indeed adequate, a better representation could be obtained (especially with large hyperpolarisations followed by a depolarisation to the sodium equilibrium potential) with twenty-fifth order kinetics. A recent analysis of the data of Cole and Moore by Woodbury (unpubl.) indicates that the adequate order of the kinetics varies with the degree of hyperpolarisation and asymptotically approaches a limiting value of 32.

H. High Potassium Effects on the K Channel Current-voltage Curve. Moore (1959) discovered that if the squid giant axon is placed in iso-osmotic potassium chloride, the steady state current versus voltage relationship for the membrane exhibits a negative slope conductance region if the membrane is hyperpolarised. Under such experimental conditions, the sodium system is largely inactivated, and the observed steady state behaviour is attributed to the late current channel. The Hodgkin-Huxley equations duplicate the results of Moore and also of Tasaki (1959) (see George and Johnson, 1961). However, Ehrenstein and Gilbert (1966) have also found a steady state negative slope conductance region in the depolarising direction under these same conditions. Thus, the current voltage relationship is semi-symmetrical about the origin. This phenomenon is not included in the Hodgkin-Huxley formalism.

Rojas and Ehrenstein (1965) found that when $n_{o,Ca} = n_{i,Ca} = 0$ and $n_{o,K} = n_{i,K} = 400$ mM/l, the steady state current voltage curve is virtually linear from -150 mV to +150 mV. Lecar, Ehrenstein Binstock and Taylor (1967) have confirmed this. They also found that addition of divalent cations externally with $n_{i,Ca} = 0$ is sufficient to induce the appearance of the negative slope conductance region. Gilbert and Ehrenstein (1969), however, have shown that a decrease in external calcium levels will shift the point of maximum steady state current in the hyperpolarising direction, indicating that the negative slope conductance region is not totally insensitive to external calcium or magnesium.

Lecar *et al.* (1967) were able to separate their current voltage curves into a linear component and a non-linear time varying component. The linear component was identified as leakage current while the non-linear component, which gave rise to the negative resistance regions in the steady state, has the kinetics of the potassium process and is decreased by the removal of divalent cations. These experiments lead to the conclusion that both the steady state negative slope conductance regions seen in high potassium are due to the potassium system.

I. The Nature of the Leakage Pathway. Hodgkin and Huxley tentatively identified the leakage current as a chloride current. This leakage component was small and in their treatment was characterized by an ohmic conductance and an equilibrium potential.

Adelman and Taylor (1961) examined the characteristics of the leakage current to test the validity of the assumptions made by Hodgkin and Huxley. They tried to eliminate the effects of the sodium and potassium currents by strong hyperpolarizations of the membrane preceding the test pulse. Following the hyperpolarizing pre-pulse, the membrane was either further hyperpolarized, or depolarized to the sodium equilibrium potential. (The effect of the strong hyperpolarization is to delay the onset of potassium current in response to a depolarization by about one-half millisecond. Depolarization to the sodium equilibrium potential, which could be changed by choline substitution for sodium, eliminated the complication of sodium currents). They found that the leakage current (current 100 μsec after the application of the voltage clamp) was not a linear function of membrane potential, and exhibited marked rectificaton . They were unable to determine the speed of development of this leakage rectification, but placed the time constant at less than 100 μsec. They found that lowering $n_{o,Ca}$ increased the leakage current. The results of changing $n_{o,Na}$, $n_{o,K}$, and $n_{o,Cl}$ led to the conclusion that less than one-tenth of the leakage current is carried by these ions and they were thus forced to postulate that the leakage current is carried primarily by the efflux of internal ions - cations for outward leakage currents, and anions for inward currents.

Fishman (1970), using voltage clamp equipment with a much

faster response time, was able to show that the marked rectification in the "leakage" current at 100 μsec is obliterated by the use of TTX and TEA^+. Thus, the non-linearity is due to ion movement through the Na and K channels. The current-voltage curve at 50 μsec is linear, and identical with the TTX-TEA modified curve at 100 μsec.

Rojas _et_ _al_. (1969, 1970) have also shown that the leakage conductance in squid giant axons is reversibly decreased by increasing the divalent (Ca^{++} or Mg^{++}) ion concentration and/or decreasing the pH of the external solution. An examination of the maximum leakage conductance as a function of pH indicated two critical points, one at a pH = 4.5 and the second at pH = 7.5. They take this evidence to indicate the involvement of carboxyl and phorphoric groups in determining the properties of the leakage conductance.

Summary. In summary, the experimental evidence strongly supports the idea that the electric field within the membrane, to which the molecular mechanisms responsible for the steady state and dynamic properties of the potassium (sodium) conductance of the squid giant axon membrane, is determined by the relative permeabilities of the resting membrane (and the consequent Goldman potential) as well as the existence of local zeta potentials, probably at both the interior and exterior borders of the membrane. With regard to the nature of these molecular mechanisms the evidence circumstantially implies that the steady state nature of the potassium (sodium) conductance is controlled by a molecular mechanism quite separate from that which controls dynamic properties. The strongest evidence for this point is that:

a) TEA^+ (TTX) is able to affect the steady state potassium (sodium) conductance without affecting the kinetic parameters of the system; and

b) Replacement of potassium (sodium) with other monovalent inorganic and organic ions has a profound effect on the steady state conductance of the late (early) current pathway, but no effect on its kinetic characteristics. These facts are further supported by the observations (Hodgkin and Huxley, 1952) that:

c) The steady state conductance of the late and early current pathways are relatively insensitive to changes in temperature (Q_{10} on the order of 1.1 - 1.3, corresponding to activation energies of the order of 2 Kcal/mole), but the kinetic parameters of both are extremely sensitive to temperature changes (Q_{10} on the order of 3, corresponding to activation energies of 19 Kcal/mole); and finally by the observation that:

d) Under the proper conditions, the internal and external perfusion of the squid giant axon with equimolar potassium chloride leads to drastic changes in the voltage dependent nature of the steady state late current pathway conductance without affecting the kinetic characteristics in the slightest.

PART II. CLASSICAL ELECTRODIFFUSION THEORIES OF MEMBRANE ELECTRICAL PROPERTIES

In Part I abundant evidence was presented pointing to the movement of ions across the plasma membrane as a prime determinant of membrane electrical properties. In Part II, I explore in some depth the numerous attempts that have been made to model these movements by the use of simple transport equations originally derived by Nernst and Planck in the 19th century. In some situations, notably the attempts to explain membrane electrical properties near equilibrium (at the resting potential), ED theory has been of great use. However, its failure to adequately deal with the data of excitable membranes far from equilibrium (as embodied in the empirical Hodgkin-Huxley formulation) is notable. This failure is a spur to develop more realistic models at the molecular level, as explored in Part III.

CHAPTER 6. CONSERVATION AND FIELD EQUATIONS

Electrodiffusion (ED) is a loose term referring to a process whereby transmembrane ion movement is determined by electric fields within the membrane and concentration gradients across the membrane. All of the derivations of the conservation equations in this chapter are to be found in Mackey and McNeel (1973), and also can be found in alternate forms in the literature. In Chapter 12 I illustrate how they arise naturally from a molecular formulation of ED theory.

Consider an ion of charge q (coulombs), mass m (g), energy u (dynes), and number density n (number per cubic centimeter) moving through an ion-permeable region of a membrane with directed velocity v_d (centimeters per second) under the influence of an externally applied electric field E (volts per centimeter) and a concentration gradient. The ion is under the influence of a force due to the electric field (qE) a force due to the concentration gradient, $\left[-d(nu)/dx\right]/n$, and a rapidly fluctuating force ($\mathcal{F}$) due to its interaction with membrane molecules. The force balance equation for the ion is

$$\frac{1}{n}\frac{\partial(mnv_d)}{\partial t} = qE + \mathcal{F} - \frac{1}{n}\frac{\partial(nu)}{\partial x}. \tag{6.1}$$

The microscopic nature of the fluctuating force is unknown, but must be characterized in order to proceed. This is done by making plausible assumptions about the average properties of the fluctuating force. If the frequency of collisions between ions of total velocity v and membrane molecules is $\nu(v)$ (collisions per second), then the deceleration experienced by an ion due to a collision is assumed to be $\nu(v)\bar{v}_d$, where $\bar{v}_d$ is the directed ionic velocity averaged over many collisions but over a time much less than ν^{-1}. So I take

$$\bar{\mathcal{F}} = -m\nu(v)\bar{v}_d \qquad (6.2)$$

If Eq. 6.1 is averaged over many collisions, the result combines with Eq. 6.2 to give

$$\frac{1}{n}\frac{\partial(mn\bar{v}_d)}{\partial t} = qE - m\nu(v)\bar{v}_d - \frac{1}{n}\frac{\partial(nu)}{\partial x} \qquad (6.3)$$

The total ionic velocity (directed plus thermal) appearing in Eq. 6.3 must not be confused with the directed ionic velocity. Even in the face of strong forces, $\bar{v}_d \ll v$; so I will alternately consider ν to be a function of ionic energy u, $\nu = \nu(u)$.

Generally speaking it is easy to see from (6.3) that in the absence of any spatial gradients and in the presence of a constant electric field the steady state terminal velocity will be $\bar{v}_{d\infty} = qE/m\nu(u)$ and it is customary to call the constant of proportionality between $\bar{v}_{d\infty}$ and qE ionic mobility μ:

$$\mu = [m\nu(u)]^{-1}$$

It is a simple matter to show that in the absence of collisions ($\nu = 0$), the ion is constantly accelerated at a value of (qE/m) and therefore a terminal velocity is not reached.

In an entirely analogous fashion suppose E = 0 for all t and that a spatial bulk energy gradient is suddenly established. Then, from (6.3), the terminal ion velocity will be $\bar{v}_{d\infty} = -[m\nu(u)n]^{-1}\partial(nu)/\partial x$ and once again the proportionality constant between terminal velocity and driving force is $\mu = [m\nu(u)]^{-1}$. For situations close enough to equilibrium that $u \simeq kT$ it is customary to define a second constant $D = kT[m\nu(u)]^{-1}$, the diffusion coefficient. Clearly, if the near equilibrium situation is not attained a natural extension of the diffusion coefficient concept is given by

$$D = u[m\nu(u)]^{-1}$$

or using the definition of mobility

$$D = \mu u$$

which becomes $D = \mu kT$ for equilibrium situations, the well known Einstein relation.

It is necessary to know how u varies with various parameters, e.g. time and electric field strength. The rate of change of the ionic energy $\dot{u}$ is the difference between the rate at which the ion gains energy from the external electric field qv_dE and from the concentration gradient $-\partial(nuv_d)/\partial x$, and the rate at which it loses energy through the rapidly fluctuating force operating during collisions $v_d \mathcal{F}$ Thus, an energy balance equation

$$\frac{1}{n}\frac{\partial(nu)}{\partial t} = qv_dE + v_d\mathcal{F} - \frac{1}{n}\frac{\partial(nuv_d)}{\partial x} \tag{6.4}$$

may be written. Knowledge of the behaviour of $\mathcal{F}$ is needed in order to proceed.

I assume, under the same averaging procedures, that the rate of loss of ionic energy is proportional to the difference between ion energy and scatterer thermal energy u, times the fractional ion energy loss per collision ξ and the collision frequency

$$\overline{v_d\mathcal{F}} = -\xi(u - u_s)\nu(u) \tag{6.5}$$

Averaging Eq. 6.4 and substituting Eq. 6.5 in the result yields

$$\frac{1}{n}\frac{\partial(nu)}{\partial t} = q\bar{v}_dE - \xi(u - u_s)\nu(u) - \frac{1}{n}\frac{\partial(nu\bar{v}_d)}{\partial x} \tag{6.6}$$

I have not yet considered the specific energy dependence of the collision frequency and energy loss per collision. It is through ν and ξ that the unique nature of the different collision processes is introduced.

For elastic collisions, if two particles interact centrally then the collision frequency may be written as a simple function of the magnitude of the relative velocity between scatterer and incident particle. If the force between ion and the fixed effective membrane molecule is given by $F_{is} = -K_{is}/r_{is}^{\alpha}$ where r_{is} is the ion-scatterer separation and α and K_{is} are constants, then (Chapman and Cowling, 1958) the collision frequency is given by $\nu(u) = \beta u^{p/2}$, where $p = (\alpha - 5)/(\alpha - 1)$ and

$$\beta = 2\pi n_s A(\alpha)\left[K_{is}(m + m_s)/mm_s\right]^{2/(\alpha-1)}$$

is a constant involving the scatterer mass m_s, number density n_s, and $A(\alpha)$ is a pure number. The average fractional ionic energy loss per collision is independent of ionic energy for elastic interactions and is given by $\xi = 2m/(m + m_s)$

Thus for dominant elastic collision processes Eq. 6.3 may be written as

$$\frac{1}{n}\frac{\partial(mn\bar{v}_d)}{\partial t} = qE - m\beta\bar{v}_d u^{p/2} - \frac{1}{n}\frac{\partial(nu)}{\partial x} \tag{6.7}$$

Eq. 6.5 becomes

$$\overline{v_d \xi} = -\zeta\beta(u - u_s)u^{p/2},$$

and Eq. 6.6. therefore takes the form

$$\frac{1}{n}\frac{\partial(nu)}{\partial t} = q\bar{v}_d E - \xi\beta(u - u_s)u^{p/2} - \frac{1}{n}\frac{\partial(nu\bar{v}_d)}{\partial x} \qquad (6.8)$$

Having obtained conservation equations for ionic momentum and energy, the final conservation equation that is needed relates to the conservation of mass (the continuity equation):

$$\frac{\partial n}{\partial t} + \frac{\partial(n\bar{v}_d)}{\partial x} = 0 \qquad (6.9)$$

In addition to the conservation equations a field equation, Poisson's equation, for the connection between electric field and ionic number densities will be essential. Poisson's equation is

$$\epsilon\frac{\partial E}{\partial x} = \sum_i q_i n_i + \sum_l q_l n_l \qquad (6.10)$$

where $\epsilon = \mathcal{K}\epsilon_o$, $\mathcal{K}$ is the membrane dielectric constant and ϵ_o is the permetivity of free space, the summation index i ranges over all mobile ions and the index l ranges over any fixed charge that may be in the membrane.

To write an equation for current density through the membrane (6.7) - (6.10) must be solved for the velocity $\bar{v}_d$. Once $\bar{v}_d$ has been obtained, the constitutive relation

$$j = qn\bar{v}_d \qquad (6.11)$$

will give the connection between $\bar{v}_d$ and ionic current density j (amperes per square centimeter).

The ionic currents due to electric field and diffusive driving forces are generally not the only current. The other current is the displacement current

$$j_d = \epsilon\frac{\partial E}{\partial t} \qquad (6.12)$$

so for time varying situations the total current, j_T, is given by

$$j_T = j_d + \sum_m j_m .$$

and j_d and j_m are both functions of x and t. However, j_T is independent of x. To demonstrate this, simply differentiate Poisson's equation with respect to t and use the continuity equation to give

$$\epsilon \frac{\partial^2 E}{\partial t \partial x} = -\sum_m \frac{\partial j_m}{\partial x} = -\frac{\partial j}{\partial x}$$

where $j = \sum_m j_m$. However,

$$\epsilon \frac{\partial^2 E}{\partial t \partial x} = \frac{\partial j_d}{\partial x}$$

so

$$\frac{\partial j_T}{\partial x} = 0$$

In dealing with equations 6.7 through 6.11 I will often find it convenient to use the dimensionless variables defined by

$$\begin{array}{lll}
u_s = kT & t_D = \delta^2/D_o & \bar{n} = (n_o+n_i)/2 \\
\nu_o = \beta u_s^{p/2} & \langle t_D \rangle = \frac{1}{p}\sum_{m=1}^{p} t_{D,m} & N = n/\bar{n} \\
\mu_o = (m\nu_o)^{-1} & & U = u/u_s \\
D_o = \mu_o kT & t_C = \nu_o^{-1} & V_d = v_d\delta/D_o \\
X = x/\delta & & \bar{E} = eE\delta/kT \\
\lambda_D^2 = kT\kappa\epsilon_o/e^2\bar{n} & T = t/\langle t_D \rangle & \varphi = eV/kT \\
 & W = \omega t_D & I = j\delta/eD_o\bar{n}
\end{array}$$

wherein n_o and n_i are the bulk extracellular and intracellular ionic concentrations of a particular ion, δ is the membrane thickness, λ_D is the Debye length, $\epsilon_o = (10^{-9}/36\pi)$ farads/metre is the permetivity of free space, κ is the membrane dielectric constant, and t_D and t_C are, respectively, characteristic times for diffusion and collision. The dimensionless time T should not be confused with the absolute temperature. The inner (intracellular) membrane border is located at $x = 0$ while the outer is at $x = \delta$ $(X=1)$.

With these definitions the conservation, field and constitutive relations may be rewritten as:

Conservation of Mass:

$$\frac{t_D}{\langle t_D \rangle}\frac{\partial N}{\partial T} + \frac{\partial (NV_d)}{\partial X} = 0 \qquad (6.12)$$

Conservation of Momentum:

$$\frac{t_C}{\langle t_D \rangle}\frac{1}{N}\frac{\partial (NV_d)}{\partial T} = Z\bar{E} - V_d U^{p/2} - \frac{1}{N}\frac{\partial (NU)}{\partial X} \qquad (6.13)$$

Conservation of Energy:

$$\frac{t_C}{\langle t_D \rangle}\frac{1}{N}\frac{\partial (NU)}{\partial T} + \frac{t_C}{t_D}\left[\frac{1}{N}\frac{\partial (NUV_d)}{\partial X} - Z\bar{E}V_d\right] = \xi U^{p/2}(1 - U) \qquad (6.14)$$

Constitutive relation for ionic current:

$$I = ZNV_d \tag{6.15}$$

Constitutive relation for displacement current:

$$I_d = \frac{\lambda_D^{\ 2}}{\delta^2} \frac{\partial \bar{E}}{\partial T} \tag{6.16}$$

Poisson's equation:

$$\frac{\lambda_D^{\ 2}}{\delta^2} \frac{\partial \bar{E}}{\partial X} = \sum_m Z_m N_m + \sum_l Z_l N_l \tag{6.17}$$

It is implicit in all of these equations that an ionic species identifying subscript m is associated with Z, N, V_d, U, I, I_d, t_D and t_C for any multiple ion situation.

The equations that are found in the membrane biophysics literature under the name of ED theory are customarily based on the unstated assumption that $t_C \ll t_D$. For this case the conservation of energy equation says that if, in addition, $t_C \ll \gtrless t_D$ then $U = 0$ (physically untenable) or $U = 1$ (i.e., ionic energy is thermal energy). Thus the conservation of momentum equation becomes

$$V_d = Z\bar{E} - \frac{1}{N}\frac{\partial N}{\partial X}$$

or, using (6.15)

$$I = Z^2 N\bar{E} - Z\frac{\partial N}{\partial X} \tag{6.18}$$

This is the equation originally derived by Nernst (1888, 1889) and Planck (1890a,b), but in dimensionless form, and traditionally referred to as the ED equation or the Nernst-Planck equation. Expressed in normal (as opposed to dimensionless) form equation 6.18 is

$$j = q^2 \mu n E - qD\frac{\partial n}{\partial x} \tag{6.19}$$

It is easy to see at this point that if it is required to know when, in a steady state (i.e., all time derivatives are zero), the single ion current carried by any particular ionic species is zero it is sufficient to note from (6.14) that U = 1 and thus (6.13) becomes

$$Z\bar{E} - \frac{1}{N}\frac{dN}{dX} = 0$$

Using $\bar{E} = -d\varphi/dX$ and integrating from X = 0 to 1 gives

$$-Z\left[\varphi(1) - \varphi(0)\right] = \int_{N_i}^{N_o} d\ln N$$

or

$$\varphi_m = \frac{1}{Z} \ln \frac{n_o}{n_i} \tag{6.19}$$

wherein $\varphi_m = \varphi(0) - \varphi(1)$. This is just the dimensionless form of the Nernst equation derived earlier in Chapter 1 from thermodynamic considerations. The value of φ_m at which the steady state ionic current is zero is the equilibrium potential and will be denoted by $\varphi_e = eV_e/kT$.

CHAPTER 7. THE STEADY STATE PROBLEM: APPROXIMATE SOLUTIONS

In this Chapter I examine simple solutions for the conservation and field equations governing electrodiffusion mediated ion movement through a membrane. Two approaches, both valid only for a true steady state, are explored. In the first approach, which assumes that the membrane thickness (δ) is small with respect to the solution Debye length (λ_D), the simplifying consequence is that the electric field is constant within the membrane. The second approximation, due to Planck, makes the opposite assumption to imply the validity of microscopic electro-neutrality.

If one of these two approximations is not made, in a true steady state situation with $(t_c/\langle t_D\rangle) \ll 1$ equation 6.18:

$$I_m = Z_m^2 N_m \bar{E} - Z_m \frac{dN_m}{dX} \tag{6.18}$$

must be combined with Poisson's equation:

$$\frac{\lambda_D^2}{\delta^2} \frac{d\bar{E}}{dX} = \sum_{m=1}^{p} Z_m N_m + \rho^* \tag{6.17}$$

wherein $\rho^* = \sum Z_e N_e$ is the net fixed charge, to give

$$N_m \frac{d^2 N_m}{dX^2} - \left(\frac{dN_m}{dX}\right)^2 - \frac{I_m}{Z_m} \frac{dN_m}{dX} = \frac{\delta^2 N_m^2}{\lambda_D^2} \left[\sum_{m=1}^{p} Z_m N_m + \rho^* \right] \tag{7.1}$$

This nonlinear equation in $N_m(X)$ is not easily solved analytically and thus there is an immediate reason to search for reasonable, simpler approximate solutions.

A. Constant Field Approximation

To effect a simplification in the solution to this problem, Goldman (1943) made the assumption that the electric field within the membrane was essentially uniform. The range of validity for this assumption may be easily derived by examining Poisson's equation (6.17) in more detail.

If $\delta^2 \ll \lambda_D^2$, then indeed the uniform, constant field approximation is not a bad one. As Cole (1965) has discussed in some detail, this condition is likely to be attained in many biological membranes and thus it seems safe to make the original Goldman assumption.

The Goldman assumption, in conjunction with Poisson's equation implies that E (= $-dV/dx$) is constant within the membrane and a simple integration with $V(x=0) = V_i$ at the interior border of the membrane and $V(x=\delta) = V_o$ at the exterior gives

$$V(x) = V_i - (V_i - V_o)X/\delta \tag{7.2}$$

Since by convention all membrane potential measurements are made with respect to the exterior solution it is consistent to take $V_o=0$ and $V_i=V_m$, the membrane potential. Thus,

$$V(x) = V_m \left[1 - x/\delta\right] \tag{7.3}$$

Integration of the ED equation 6.19, rewritten as

$$-\frac{j}{qD} = \frac{dn}{dx} + \frac{qn}{KT}\frac{dV}{dx} \tag{7.4}$$

with the aid of the Einstein relation (Chapter 6), is easily effected by multiplying (7.4) by the integrating factor $\exp(qV/KT)$ to give

$$-(j/qD)\exp(Z\varphi)dx = d\left[n(x)\exp(Z\varphi)\right] \tag{7.5}$$

where I have used $\varphi(x) = eV(x)/KT)$, or

$$j = \frac{qD}{\int_0^\delta \exp(Z\varphi)dX}\left[n_i\exp(Z\varphi_m) - n_o\right] \tag{7.6}$$

wherein $n(x=0) = n_i$, $n(x=\delta) = n_o$. In arriving at (7.6), use must be made of the fact that $j \neq j(x)$ from the conservation of mass equation when $\dot{n} = 0$.

Now from (7.3)

$$\int_0^{\delta} \exp(Z\varphi)\,dx = \frac{\delta}{Z\varphi_m}\left[\exp(Z\varphi_m) - 1\right] \tag{7.7}$$

so combining this with equation 7.6 gives an equation for ionic current density in terms of membrane potential and intra- and extra-cellular ionic concentrations:

$$j = \frac{qZ\varphi_m D}{\delta}\left[\frac{n_i \exp(Z\varphi_m) - n_o}{\exp(Z\varphi_m) - 1}\right] \tag{7.8}$$

It is clear from (7.8) that for $Z\varphi_m >> 0$,

$$j \simeq qZ\varphi_m D n_i/\delta \tag{7.9a}$$

while for $Z\varphi_m << 0$,

$$j \simeq qZ\varphi_m D n_o/\delta \tag{7.9b}$$

Further, $j = 0$ when $n_i \exp(Z\varphi_m) = n_o$, i.e., when $\varphi_m = \varphi_e$; and at $\varphi_m = 0$,

$$j = qD(n_i - n_o)/\delta \tag{7.9c}$$

Thus, the qualitative forms of the current versus membrane potential characteristics predicted by the constant field treatment of the electrodiffusion equation will be as illustrated in Figure 7.1a. In that figure I have plotted $j' = \delta j/qZD$ versus φ_m and it is clear that the two graphs cover the four possible cases, i.e. i) $Z > 0$, $n_o < n_i$; ii) $Z < 0$, $n_o > n_i$; iii) $Z > 0$, $n_o > n_i$; and iv) $Z < 0$, $n_o < n_i$.

How does the steady state behaviour depicted in Figure 7.1a compare with actual steady state single ion currents in biological membranes? The K^+ and Na^+ currents described in Chapter 4 for the squid giant axon membrane offer a convenient set of data for comparison with theory. With respect to potassium current there is qualitative agreement between theory and experiment. Measured potassium currents display the same type of rectification predicted by equation 7.8, although the predicted rectification ratio

$$\frac{j(Z\varphi_m >> 0)}{j(Z\varphi_m << 0)} = \frac{n_i}{n_o}$$

is of the order of 20 while the experimentally determined one is about 100 (Cole and Moore, 1960).

A further discrepancy between theory and experimental findings

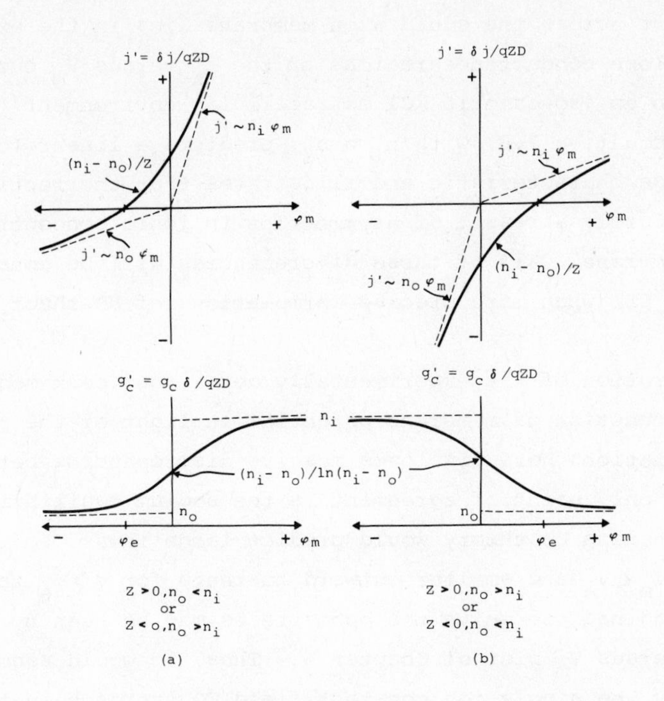

Figure 7.1 A schematic representation of the expected current-membrane potential and chord conductance-membrane potential relations for electrodiffusion mediated transmembrane ion movement.

for K^+ movement across the squid axon membrane lies in the existence of negative slope conductance regions on the j_K versus V_m curve when the axon is in an iso-osmotic KCl extracellular environment (see Chapter 5). Equation 7.8, with $n_i = n_o$, predicts a linear (ohmic) current-voltage characteristic and illustrates that the rectification predicted is purely a result of asymmetries in ionic concentrations across the membrane. All of these discrepancies will be considered again in Part III when more precise formulations of ED theory are considered.

Consideration of the experimentally determined peak sodium current as a function of membrane potential in light of the theoretical considerations here discloses massive discrepancies between the two - the only point of agreement is the sodium equilibrium potential. Whereas ED theory would predict large inward sodium currents for $V_m < V_e$ and smaller outward currents for $V > V_e$ the experimental findings are quite the opposite as may be seen by examining the s_∞ versus V_m plot of Chapter 4. Thus, it would seem that most certainly the single ion constant field ED approach in the form used here contains serious deficiencies in modeling transmembrane sodium movement.

As used by Hodgkin and Huxley (Chapter 4) the defining relation for the chord conductance $g_c(V_m, V_e)$ is

$$j(V_m, V_e) = (V_m - V_e) g_c(V_m, V_e) \tag{7.10}$$

Another measure of conductance is the slope conductance, $g_s(V_m, V_e) = (dj/dV_m)$. Obviously from (7.10) g_s and g_c are related by

$$g_s = g_c + (V_m - V_e) dg_c/dV_m \tag{7.11}$$

Both g_s and g_c are used to describe the behaviour of ionic systems, as has been noted earlier in descriptions of experimental data. How do $g_c(V_m, V_c)$ and $g_s(V_m; V_e)$, as predicted by (7.8), behave? From (7.8) and (7.10),

$$g_c(\varphi_m, \varphi_e) = \frac{qZD\varphi_m}{\delta(\varphi_m - \varphi_e)} \left[\frac{n_i \exp(Z\varphi_m) - n_o}{\exp(A\varphi_m) - 1} \right] \tag{7.12}$$

and it is easy to show that

$$g_c(Z\varphi_m >> Z\varphi_e) \simeq qZDn_i/\delta \tag{7.13a}$$

while

$$g_c(Z\varphi_m \ll Z\varphi_e) \simeq qZDn_o/\delta \tag{7.13b}$$

Further,

$$g_c(\varphi_m=0) = qZD(n_i-n_o)/\delta \ln(n_i/n_o) \tag{7.13c}$$

and

$$g_c(\varphi_m=\varphi_e) = qZDn_in_o\ln(n_i/n_o)/\delta(n_i-n_o) \tag{7.13d}$$

to allow a qualitative sketch of g_c versus φ_m to be drawn as in Figure 7.1b for the four possible cases.

If it is not possible to separate total steady state membrane ionic current into its components, one is faced with interpreting much different data than I have been considering to this point in this chapter. A simple, but realistic, situation for a squid giant axon would be to assume that the total steady state ionic current is carried by Na^+, K^+ and Cl^-. Rewrite equation 7.8 in the form

$$j = qZ\varphi_m P\left[\frac{n_i\exp(Z\varphi_m) - n_o}{\exp(Z\varphi_m) - 1}\right] \tag{7.14}$$

where $(D/\delta) = P$, the <u>permeability</u> of the membrane, so

$$j_{Na} = e\varphi_m P_{Na}\left[\frac{n_{i,Na}\exp\varphi_m - n_{o,Na}}{\exp\varphi_m - 1}\right] \tag{7.15a}$$

$$j_K = e\varphi_m P_K\left[\frac{n_{i,K}\exp\varphi_m - n_{o,K}}{\exp\varphi_m - 1}\right] \tag{7.15b}$$

and (remembering that $Z_{Cl} = -1$)

$$j_{Cl} = e\varphi_m P_{Cl}\left[\frac{n_{o,Cl}\exp\varphi_m - n_{i,Cl}}{\exp\varphi_m - 1}\right] \tag{7.15c}$$

Thus the total ionic current through the membrane is $j = j_{Na}+j_K+j_{Cl}$ or

$$j = e\varphi_m\left[\frac{A\exp\varphi_m - B}{\exp\varphi_m - 1}\right] \tag{7.16}$$

wherein

$$A = P_Kn_{i,K} + P_{Na}n_{i,Na} + P_{Cl}n_{o,Cl} \tag{7.16a}$$

and

$$B = P_Kn_{o,K} + P_{Na}n_{o,Na} + P_{Cl}n_{i,Cl} \tag{7.16b}$$

From equation 7.16 it is easy to see that the total current j will be zero at a value of φ_m given by

$$\varphi_R = \ln(A/B) \qquad (7.17)$$

or

$$V_R = \frac{KT}{e} \ln\left[\frac{P_K n_{i,K} + P_{Na} n_{i,Na} + P_{Cl} n_{o,Cl}}{P_K n_{o,K} + P_{Na} n_{o,Na} + P_{Cl} n_{i,Cl}}\right] \qquad (7.18)$$

Operationally, the potential V_R at which the total experimentally measured ionic membrane current is zero is the cell resting potential. Equation 7.18 (or 7.17) is known as the Hodgkin-Katz (1949) equation and seems to give good estimates of V_R when independent estimates of A and B are available. Further, with these estimates (7.16) accurately predicts the form of the total membrane current - voltage characteristic in the absence of excitation.

The resting potential is not related in a simple fashion to the single ion equilibrium potentials, and it should be specifically noted that $V_R \neq V_{e,Na} + V_{e,K} + V_{e,Cl}$. It is a straightforward exercise to analyze the $j(V_m, V_r)$ characteristics for a multiple ion system described by equation 7.16. Further it is trivial to generalize this treatment to an arbitrary number of cation and anion species when the absolute value of their valences are all equal. However, if this restriction is not met the problem is much more difficult and has, to my knowledge, not been solved.

B. The Microscopic Electroneutrality Approximation

The original solutions of the Nernst-Planck equation were for 'thick' liquid junctions in which $\delta^2 >> \lambda_D^2$, thereby implying from Poisson's equation that

$$\bar{\rho} = \frac{\lambda_D^2}{\delta^2} \frac{d\bar{E}}{dx} \sim 0$$

and giving rise to the microscopic electroneutrality assumption $\rho = \sum qn = 0$.

Planck presented the first solution to this problem for a 1-1 univalent electrolyte, but it is easily generalized to $Z_+ = -Z_- = Z$. In this case the ED equations 6.19 for cation and anion become

$$-\frac{j_+}{qD_+} = \frac{dn_+}{dx} + Zn_+ \frac{d\varphi}{dx} \qquad (7.19a)$$

and

$$\frac{j_-}{qD_-} = \frac{dn_-}{dx} - Zn_-\frac{d\varphi}{dx} \tag{7.19b}$$

respectively. Since this is 1-1 electrolyte $n_+ = n_- = n$, and adding and subtracting equations 7.19a,b gives

$$-\frac{1}{q}\left(\frac{j_+}{D_+} - \frac{j_-}{D_-}\right) = 2\frac{dn}{dx} \tag{7.20a}$$

and

$$-\frac{1}{q}\left(\frac{j_+}{D_+} + \frac{j_-}{D_-}\right) = 2Zn\frac{d\varphi}{dx} \tag{7.20b}$$

respectively. Integrating (7.20a) across the membrane gives

$$n(x) = n_i - \frac{1}{2q}\left(\frac{j_+}{D_+} - \frac{j_-}{D_-}\right)x \tag{7.21}$$

where $n_i = n(x{=}0)$ as before. Further, $n(x{=}\delta) = n_o$ so

$$\frac{j_+}{D_+} - \frac{j_-}{D_-} = -\frac{2q(n_o - n_i)}{\delta} \tag{7.22}$$

and thus

$$n(x) = n_i + (n_o - n_i)x/\delta \tag{7.23}$$

Rewriting (7.20b) and using (7.23) gives

$$\frac{d\varphi}{dx} = -\frac{1}{2Zq}\left(\frac{j_+}{D_+} + \frac{j_-}{D_-}\right)\frac{1}{n_i + (n_o - n_i)x/\delta}$$

or after integrating

$$\varphi_m = \frac{\delta}{2Zq}\left(\frac{j_+}{D_+} + \frac{j_-}{D_-}\right)\frac{\ln(n_o/n_i)}{n_o - n_i} \tag{7.24}$$

Thus

$$\frac{j_+}{D_+} + \frac{j_-}{D_-} = \frac{2Zq\varphi_m(n_o - n_i)}{\delta\ \ln(n_o/n_i)} \tag{7.25}$$

and j_+ and j_- may now be obtained directly.

Adding equations 7.22 and 7.25 gives

$$j_+ = 2qD_+(n_o - n_i)(\varphi_m - \varphi_e)/\delta\,\varphi_e \tag{7.26}$$

while subtraction yields

$$j_- = 2qD_-(n_o - n_i)(\varphi_m + \varphi_e)/\delta\,\varphi_e \tag{7.27}$$

Thus each single ion current consists of a diffusion current proportional to (n_o-n_i) and an ohmic current. The total current, $j = j_+ + j_-$, will be zero when $\varphi_m = \varphi_D$, the diffusion potential, which is given by

$$\varphi_D = \frac{D_+ - D_-}{D_+ + D_-} \varphi_e \tag{7.28}$$

PROBLEMS

7.1. Derive an expression for g_s from the Goldman equation. What are $g_s(\varphi_m = \varphi_e)$, $g_s(\varphi_m = 0)$, $\lim_{z\varphi_m \to \pm\infty} g_s$?

7.2. By marking an intracellular ion one can measure a one way current, I^{out}, carried by that ionic species out of the cell. In a similar fashion, ions in the extracellular fluid can be marked to give a one way current, I^{in}. Define the flux ratio, or one way current ratio, by $R_F = | I^{in}/I^{out} |$. Obtain an expression for R_F from (7.8). Under what situations is your result valid? Is it valid for more general conditions than (7.8)?

7.3. Generalize equations 7.16 and 7.17 to the case of l cation species and m anion species, where $|z_l| = |z_m| = z$ for all l and m.

7.4. Derive an expression for the membrane chord conductance and slope conductance from (7.16). Express your answer in terms of A and B, and characterize the behavior of g_c and g_s analytically and graphically as in the single ion case.

7.5. A frog sartorius muscle cell has the following ionic concentrations: $n_{i,Na} = 12$ mM/L; $n_{o,Na} = 145$; $n_{i,Cl} = 4$; $n_{o,Cl} = 120$; $n_{i,K} = 155$. Plot V_R against ln $n_{o,K}$ for $n_{o,K}$ between 1 and 300 mM/L. Take $P_{Na}/P_K = 0.02$ and $P_{Cl}/P_K = 1$, $T = 300^{\circ}K$. Compare the result with the potassium equilibrium potential calculated from the Nernst equation for the same range of external potassium concentrations.

7.6. For the same conditions as in problem 7.5, $n_{o,K} = 2.5$ mM/L, and $P_K = 2 \times 10^{-6}$ cm/sec, plot $j = j_K + j_{Na} + j_{Cl}$ as a function of V_m between -150 mV and 150 mV. Compare the result with j_K versus V_m over the same range, and also examine the chord and slope conductance variations over the same membrane potential range.

7.7. I have been doing experiments on an artificial membrane system (c.f. Problem 1.1) modified with 10 picograms LTD/cm^2 of membrane (1 picogram = 10^{-12} grams) and find that currents through the LTD modified membrane agree with electrodiffusion theory predictions.

With 10^{-3} M/L KCl on each side of the membrane, g_c = 1 mmho/cm^2 = 10^{-3} mho/cm^2. Assuming that I have maintained KCl at a concentration of 10^{-3} M/L on the "outside" and 10^{-1} M/L on the "inside":

a. Sketch the membrane potassium current density (j_K; amps/cm^2) as a function of membrane potential (V_m, millivolts) assuming that K^+ are the only charge carriers going through the membrane "channels" formed by the protein additive. Carefully indicate on your sketch the numerical values of:

i. j_K at $V_m = 0$

ii. $V_{e,K}$ (the potassium equilibrium potential)

iii. The limiting asymptotic relation between j_K and V_m for large positive and negative values of V_m.

b. Sketch the membrane potassium chord conductance (g_c, mhos/cm^2) as a function of membrane potential, V_m based on the results of section "a" above. Carefully indicate on your sketch the numerical value of:

i. g_c at $V_m = 0$

ii. g_c at $V_m = V_{e,K}$

iii. The limiting asymptotic values of g_c for large positive and negative values of V_m.

7.8. Generalize the treatment of the microscopic electroneutrality two ion case to $Z_+n_+ = -Z_-n_- = Zn$.

7.9. Generalize the microscopic electroneutrality two ion case to include l cation species, valence Z_+, and m anion species, valence Z_-. Show that the ionic concentrations are non-linear across the membrane and that the ionic currents are now non-linear functions of membrane potential.

7.10. Schlögl (1954) has solved the general problem of electrolyte mixtures with an arbitrary number of ionic valence types. Solve the problem for yourself.

CHAPTER 8. ACTIVE TRANSPORT AND THE MAINTENANCE OF TRANSMEMBRANE IONIC DISTRIBUTIONS.

In all the discussion to this point I have not considered the question of how biological cells are able to maintain asymmetric transmembrane ion concentrations. In most cells there is a high external and a low internal concentration of Na^+; the situation is reversed for K^+. Taking Na^+ as an example, it would be expected to diffuse into a cell because of the concentration gradient. Further, there should be an influx of Na^+ because of the membrane electric field. From simple thermodynamic concepts this process should continue until external and internal Na^+ concentrations are equal.

Thus given the forces acting on ions, $n_{i,Na}$ should increase as a function of time. At what rate should this occur? Consider a single cell (e.g. muscle fibre) of length l and diameter d. Let the influx of Na^+ be denoted by M_{Na}^{in} (dimensions: moles/cm^2 sec), so as a first approximation

$$\frac{\pi d^2 l}{4} \frac{dn_{i,Na}}{dt} = \pi d l M_{Na}^{in}$$

or

$$\frac{dn_{i,Na}}{dt} = \frac{4}{d} M_{Na}^{in}$$

The diameter of a single cell from the semi-tendinosis muscle of the frog is about 100μ, $M_{Na}^{in} \simeq 4 \times 10^{-12}$ M/cm^2 sec and $n_{i,Na} \simeq 10$mM/L = 10^{-5} M/cm^3 (Hodgkin and Horowicz, 1959) so

$$\frac{dn_{i,Na}}{dt} \simeq 2 \times 10^{-9} \text{ M/cm}^3\text{-sec}$$

Thus in a period of roughly 4.5 hours (1.8×10^4 sec) the internal sodium concentration of this muscle fibre would double.

The above discussion forces the conclusion that there must be some mechanism, not previously considered, capable of moving Na^+ out of cells against both diffusive forces and electric forces. This, in turn, implies the existence of some energy source. The energy source is the metabolic energy pool of the cell, derived from oxidative phosphorylation. Ion transport that requires this supply of metabolic energy is usually referred to as active transport. Active transport is a poorly understood transport mechanism in membrane physiology. Most of the available theoretical discussion concerning its characteristics is highly speculative. The molecular mechanism(s) whereby active transport systems function are, contrary to the claims of many investigators in the field, shrouded in mystery.

Below I discuss some of the known characteristics of active ion transport and look at the consequences of incorporating active transport fluxes into the electro-diffusion approach to membrane electrical properties. Finally I consider the time dependent alterations in $n_{i,K}$, $n_{i,Na}$, and $n_{i,Cl}$ during repetitive firing in a fibre whose active transport mechanisms are operating normally.

A. General Characteristics of Active Transport Systems.

Active transport processes are best studied by looking at one way ion movements across the membrane, e.g. the efflux of Na^+. Electrical measurements thus become unsatisfactory, because they measure net ion movement, and recourse must be taken to looking at the movements of ions "tagged" in some way. The use of radioactive isotopes is an obvious choice. Because of the difficulties in radioactive counting procedures, it is advantageous to deal with as large a cell as possible in carrying out these investigations. Single muscle cells (Hodgkin and Horowicz, 1959) and single giant axons from Sepia (lobster) and Loligo (squid) (Hodgkin and Keynes, 1955) have yielded the characteristics of active transport that I consider first.

Hodgkin and Keynes measured the molar influx and efflux of both Na^+ and K^+ across the axonal membrane and examined the effects of temperature and ionic changes and various inhibitors of oxidative phosphorylation on these fluxes. In the following summary of their results I use the symbols M_{Na}^{in}, M_{K}^{in}, M_{Na}^{out}, and M_{K}^{out} to denote these fluxes. They found that:

1. $M_{Na}^{out} \simeq M_{K}^{in}$
2. M_{Na}^{out} was directly proportional to $n_{i,Na}$ for $n_{i,Na} \leq$ 130mM/L
3. When metabolic inhibitors such as dinitrophenol (DNP), sodium azide (NaN_3) and cyanide (CN) were applied to the axon there was a marked, but reversible, decrease in both M_{Na}^{out}, and M_{K}^{in}. The drop in M_{Na}^{out} was approximately equal to the decrease in M_{K}^{in}. The fluxes M_{Na}^{in} and M_{K}^{out} were unaffected by these metabolic inhibitors.
4. A decrease in $n_{o,K}$ to zero decreased M_{Na}^{out} but did not totally abolish it. An elevation of $n_{o,K}$ gave corresponding increases in M_{Na}^{out}.
5. When temperature was varied it was discovered that M_{Na}^{out} and M_{K}^{in} were highly temperature sensitive, with Q_{10}'s of 3.3. However M_{Na}^{in} and M_{K}^{out} were quite temperature insensitive, with Q_{10} values of 1.4 and 1.1 respectively.

These results are qualitatively identical with those obtained by Hodgkin and Horowicz using muscle cells. They have led to the concept that active transport mechanisms are intimately dependent on cellular sources of metabolic energy, although it is not completely clear what sequence of molecular events connect the oxidative phosphorylation process and active ion movement. These results have further led to the view that the active transport of Na^+ and K^+ are somehow linked together. Again, the nature of the linkage is unclear. The roughly 1:1 correspondence between changes in M_{Na}^{out} and M_{K}^{in} initially led to the erroneous concept (see below) that on the average one K^+ is transported into the cell for every Na^+ ejected.

It has been noted by a number of investigators that the active transport fluxes of Na^+ and K^+ increase greatly in excitable tissue upon stimulation. There are two factors, noted by Hodgkin and Keynes, that might contribute to this increased M_{Na}^{out} and M_{K}^{in}. During prolonged excitation $n_{i,Na}$ should increase as should $n_{o,K}$ and both of these factors would lead to increased active transport fluxes. Further experimental work on frog skeletal muscle (Keynes and Swan,

1959; Mullins and Frumento, 1963) indicates that the relation between M_{Na}^{out} and $n_{i,Na}$, may not be the simple linear relationship reported by Hodgkin and Horowicz. These investigators have reported $M_{Na}^{out} \propto n_{i,Na}^{3}$, and their observation has been confirmed by many other workers in muscle but not axons. The discrepancy between the two reported dependencies of M_{Na}^{out} on $n_{i,Na}$ may be simply due to the restricted range of $n_{i,Na}$ investigated by Hodgkin and Horowicz. In any event, a situation where $M_{Na}^{out} \propto n_{i,Na}^{3}$ would give a more sensitive control of active sodium extrusion than would a simple linear relationship.

Evidence of a further factor influencing M_{Na}^{out} is given by Horowicz and Gerber (1959a,b) who present evidence indicating that M_{Na}^{out} may be sensitive to membrane potential. Techniques designed to depolarize the membrane (take V_m closer to zero) appear to have the effect of increasing M_{Na}^{out}. More recent data of Brinley and Mullins (1974) makes this questionable, however, and it would seem to remain unresolved.

Lastly we note that in muscle there is now considerable evidence (Mullins and Noda, 1963; Adrian and Slayman, 1966; Mullins and Awad, 1965; Frumento, 1965) that the 1:1 correspondence between Na^+ efflux and K^+ influx does not always hold. It has been found that about 3 Na^+ are actively extruded from the cell for every 1 K^+ taken up in the above quoted studies on muscle. Further, Senft (1967) has concluded that in lobster axon there must also be a non 1:1 active Na-K coupling, and his speculation is confirmed by Sjodin and Beauge (1967, 1968) who found a 2:1 Na-K coupling in the squid giant axon. Indeed, as Sjodin and Beauge point out there was evidence for this in the work of Hodgkin and Keynes (1955). They noted that the application of metabolic inhibitors reduced M_K^{in} from 22 to 3 pM/cm^2-sec while M_{Na}^{out} was reduced from 39 to 3 pM/cm^2-sec. Thus taking the portions of M_{Na}^{out} and M_K^{in} due to active transport as 36 and 19 pM/cm^2-sec respectively this yields a figure of 36:19 = 2.1:1 as the coupling ratio. The consequence of having a membrane located active transport system that moves more Na^+ out of a cell than K^+ in a given period of time is that the mechanism will play a direct role in separating charge and thus affecting the membrane potential. A 1:1 electroneutral pump would have no direct effect on the membrane potential. The evidence for a non 1:1 electrogenic sodium pump has been extensively reviewed by Thomas (1972), and he concludes that there is little if any evidence for the existence of an electro-neutral pump in any vertebrate and invertebrate neurons or muscle cells. Evidence from a number of tissues indicates a coupling ratio of 2:1 or 3:2, and data of Sjodin and Beauge (1968) would indicate a variable coupling ratio that increases as $n_{i,Na}$ increases.

A further item of interest with respect to the operation of the pump was explored by Sjodin and Beauge (1967,1968). They examined the effects on M_{Na}^{out} when external K^+ was replaced by either Rb^+ or Cs^+. Both Rb^+ and Cs^+ replacement of external K^+ decreased the active transport of Na^+ out of the axon, and from their data they were able to derive a relative selectivity of the active transport mechanism that is K:Rb:Cs = 1:0.84:0.22.

In a series of experiments by Caldwell _et al_. (1960a,b) clues to the critical characteristics of the energy supply for active transport processes were obtained. Using squid giant axons they determined that the predominant high energy phosphate compounds in the axoplasm were adenosine tri-phosphate (ATP), arginine phosphate (Arg P), and orthophosphate. The effect of the metabolic inhibitors cyanide, DNP, and sodium azide is to drastically decrease the axoplasmic concentrations of ATP and Arg P, thereby implicating these two compounds as active transport energy sources.

In squid giant axons that have been metabolically poisoned, the external application of ATP and/or Arg P has no effect on active ion fluxes. However, the intracellular application of ATP or Arg P again stimulates the extrusion of Na. The M_{Na}^{out} stimulated by Arg P injection is sensitive to $n_{o,K}$ as in normal axons, but the M_{Na}^{out} stimulated by ATP injection is unresponsive to changes in $n_{o,K}$. The intracellular application of Arg P will further stimulate M_K^{in} in the poisoned axon, while ATP is without effect. All of these procedures leave M_{Na}^{in} and M_{Na}^{out} unaffected.

These results lead to the conclusion that the active M_K^{in} is dependent on the presence of intracellular Arg P while M_{Na}^{out} has two components. One depends on ATP and the second, K^+ sensitive one, depends on Arg P. The "linkage" between the K^+ sensitive M_{Na}^{out} and M_K^{in} is dependent on the presence of Arg P.

Ouabain and the cardiac glycosides, e.g. digitalis, have been known for some time to inhibit the active transport of Na^+ and K^+ in excitable tissue (Glynn, 1964). Experiments by Caldwell (1960) and Caldwell and Keynes (1959) have shown that these compounds have no effect on axoplasmic ATP or Arg P concentrations. They act to inhibit active ion transport other than by removing the energy source. In contrast to the metabolic poisons like DNP they are effective in disrupting active transport only if applied on the outside of the cell. Apparently some portion of the active transport mechanism sensitive to them is accessible only from the outside of the membrane.

Extensive work on a number of inexcitable tissues has confirmed the existence of active ion transport mechanisms qualitatively identical in operation to the mechanism in excitable tissue discussed above. Skou (1965) reviews the evidence from many tissues indicating that their active transport mechanism: 1) Is located in the cell membrane; 2) Has an affinity for Na^+ over K^+ at the inner membrane border; 3) Has an affinity for K^+ over Na^+ at the outer membrane border; 4) Must contain an enzyme able to convert the energy released by ATP hydrolysis into cation movement; and 5) Operates at a rate dependent on $n_{i,Na}$ and $n_{o,K}$.

A series of experiments by Skou ultimately led to the identification of a membrane bound ATP hydrolyzing enzyme system that requires Na^+ and K^+ for maximal activity, and conforms to the above requirements for a transport system. In addition, its behaviour in the presence of the cardiac glycosides, and in response to Na^+ and K^+ is quite similar to the analogous behaviour of the intact transport system. In what follows I briefly summarize the properties of this enzyme system and indicate its possible relationship to intact active transport systems, closely following Skou (1965).

This enzyme system (denoted by TS) hydrolyzes ATP to ADP and phosphate (Pi). For the enzyme to be active, Mg^{++} is required and maximal activity is attained only in the presence of Na^+, K^+, and Mg^{++}. In an experiment in which Mg^{++} and TS are present, the addition of Na^+ to the medium leads to an increase in TS activity. The addition of K^+, Rb^+, NH_4^+, or Cs^+, however, has a less pronounced effect on TS activity.

If the same experiment is carried out with Mg^{++}, Na^+, and TS present, and then other monovalent cations are added, considerable increases in TS activity are noted. The ions NH_4^+, K^+, Cs^+, and Li^+ are all effective in this regard with NH_4^+ being the most effective and Li^+ the least. Apparently TS is able to bind all of these ions and it has been found that the binding affinity sequence, from high to low, is K^+, Rb^+, NH_4^+, Cs^+ and Li^+ in contrast to the activity sequence.

The above experiment with Mg^{++}, Na^+, and TS present and variable amounts of K^+ show that TS activity increases to a maximum with increasing K^+ concentration and then declines to the level induced by Mg^{++} alone. A kinetic analysis of this data strongly supports the idea that TS contains two binding sites. One has an affinity for Na^+ that is about 8 times that for K^+, while the second has a K^+ affinity greater than that for Na^+. Evidence indicates that, in the intact cell, the high sodium affinity site is located at the inner membrane border while the high potassium affinity site is at the outer membrane border.

Experiments on the TS system in conjunction with various cardiac glycosides demonstrate that Mg^{++} induced activity is unaffected by these compounds, but that the Na^+ and K^+ induced TS activity is decreased. Further kinetic experiments indicate that the cardiac glycosides compete with K^+ for a binding site on TS, thus offering insight into the fact that the glycosides are effective at the outer border of the intact cell membrane. Experiments with various glycosides indicate a strict correlation between their competition for the K^+ binding site and the ability to interfere with active transport in the intact cell.

B. The Consequences of Including Active Transport of Na^+ and K^+ in Steady State Electrodiffusion.

In this section I want to examine the consequences of including currents due to the active transport of Na^+ and K^+ in the electrodiffusion formulation of membrane transport. The treatment is after Woodbury (1965). Write the current carried by the active transport of Na^+ as

$$j_{Na}^{AT} = eJn_{i,Na} \quad (8.1)$$

and that due to K^+ by

$$j_{K}^{AT} = eJn_{i,Na}/r \quad (8.2)$$

In (8.1) and (8.2) J is a constant with the dimensions of cm/sec (the same as a permeability), and r is the coupling coefficient that allows for other than 1:1 coupling of Na^+ and K^+ fluxes, e.g. r = 1 for 1:1; r = 3 for 3:1 coupling. Note that although equations 8.1 and 8.2 incorporate most of the active transport characteristics discussed above, they neglect the possibility of a non-linear relation between M_{Na}^{out} and $n_{i,Na}$, as reported for frog muscle. The inclusion of this property is a simple matter and is left as an exercise.

To avoid a second mathematical complexity I will use equation 7.6 to describe each of the ionic currents, define a different permeability by

$$\bar{P}(\varphi_m) = D\exp(Z\varphi_m) / \int_0^{\delta} \exp(Z\varphi)dx \quad (8.3)$$

and assume $\bar{P}$ to be essentially independent of the membrane potential (which is nonsense for some situations). Thus the expressions analogous to (7.15a-c) become

$$j_{Na} = e\bar{P}_{Na}\left[n_{i,Na} - n_{o,Na}\exp(-\varphi_m)\right] \quad (8.4a)$$

$$j_{K} = e\bar{P}_{K}\left[n_{i,K} - n_{o,K}\exp(-\varphi_m)\right] \quad (8.4b)$$

and

$$j_{Cl} = e\bar{P}_{Cl}\left[n_{i,Cl} - n_{o,Cl}\exp\varphi_m\right] \quad (8.4c)$$

With equations 8.1 and 8.2 in conjunction with 8.4 I write expressions for the net sodium, potassium and chloride currents.

$$j_{Na}^{net} = e\bar{P}_{Na}\left[n_{i,Na} - n_{o,Na}\exp(-\varphi_m)\right] + eJn_{i,Na} \quad (8.5a)$$

$$j_{K}^{net} = e\bar{P}_{K}\left[n_{i,K} - n_{o,K}\exp(-\varphi_m)\right] - (eJ/r)n_{i,Na} \quad (8.5b)$$

$$j_{Cl} = e\bar{P}_{Cl}\left[n_{i,Cl} - n_{o,Cl}\exp\varphi_m\right] \quad (8.5c)$$

I want to look at the situation when each permeable ion is in equilibrium across the membrane, so $j_{Na}^{net} = j_{K}^{net} = j_{Cl}^{net} = 0$. Further I assume that the only known quantities are the permeabilities and the external ion concentration, and ask for the equilibrium values of V_m and internal ion concentrations.

Thus there are four unknown quantities, and (8.5) in conjunction with the equilibrium conditions constitutes only three equations. To remedy this, introduce the reasonable assumption of bulk intracellular charge neutrality given by

$$n_{i,Na} + n_{i,K} - n_{i,Cl} - Z_A n_{i,A} = 0 \quad (8.6)$$

where A denotes the predominant intracellular organic anion, and Z_A is its valence. The total number (A_T) of organic anions in the cell is fixed because the cell membrane is impermeable to them. Thus if the cellular volume is V (not necessarily constant) $n_{i,A} = A_T/V$, and (8.6) becomes

$$n_{i,Na} + n_{i,K} - n_{i,Cl} - (ZA_T/V) = 0 \quad (8.7)$$

and a new variable, V has been introduced. ($Z_A \simeq 1$ and A_T is known through chemical analysis).

To obtain an additional equation, further assume that the intra- and extracellular osmotic pressures are equal (again, quite reasonable in light of the high water permeability of the membrane):

$$n_{i,Na} + n_{i,K} + n_{i,Cl} + (ZA_T/V) = n_{o,Na} + n_{o,K} + n_{o,Cl} \quad (8.8)$$

There are now five equations in the five unknowns $n_{i,Na}$, $n_{i,K}$, $n_{i,Cl}$, φ_m and V with the other quantities available from data.

For notational convenience, define a new variable by $\alpha = \exp(-\varphi_m)$ so equations 8.5 become

$$j_{Na}^{net} = e\bar{P}_{Na}(n_{i,Na} - \alpha n_{o,Na}) + eJn_{i,Na} \quad (8.9a)$$

$$j_{K}^{net} = e\bar{P}_{K}(n_{i,K} - \alpha n_{o,K}) - (eJ/r)n_{i,Na} \quad (8.9b)$$

and

$$j_{Cl}^{net} = e\bar{P}_{Cl}(n_{i,Cl} - \alpha^{-1} n_{o,Cl}) \quad (8.9c)$$

Invoking the true steady state condition on j_{Na}^{net}, j_{K}^{net}, and j_{Cl}^{net} gives

$$\bar{P}_{Na}(n_{i,Na} - \alpha n_{o,Na}) + Jn_{i,Na} = 0 \quad (8.10)$$

$$\bar{P}_{K}(n_{i,K} - \alpha n_{o,K}) - (J/r)n_{i,Na} = 0 \quad (8.11)$$

and

$$\alpha n_{i,Cl} - n_{o,Cl} = 0 \quad (8.12)$$

Solving (8.10) for $n_{i,Na}$ gives

$$n_{i,Na} = \bar{P}_{Na} n_{o,Na} \alpha / (\bar{P}_{Na} + J) \tag{8.13}$$

and adding (8.7) and (8.8) gives

$$n_{i,K} = (n_{o,Na} + n_{o,K} + n_{o,Cl})/2 - n_{i,Na} \tag{8.14}$$

Substituting (8.14) into (8.11) yields a second expression for $n_{i,Na}$ involving all known quantities except for α:

$$n_{i,Na} = \bar{P}_K \left[n_{o,Na} + (1-2\alpha) n_{o,K} + n_{o,Cl} \right] / 2 \left[\bar{P}_K + (J/r) \right] \tag{8.15}$$

Equating (8.13) and (8.15) gives

$$\alpha \left[\frac{n_{o,Na}}{1+(J/\bar{P}_{Na})} + \frac{n_{o,K}}{(1+J/rP_K)} \right] = \frac{n_{o,Na}+n_{o,K}+n_{o,Cl}}{2(1 + J/rP_K)}$$

or, using $\varphi_m = -\ln\alpha$

$$\varphi_m = -\ln \left[\frac{\bar{P}_K(\bar{P}_{Na}+J)(n_{o,Na} + n_{o,K} + n_{o,Cl})}{2\left[\bar{P}_{Na}(\bar{P}_K+J/r) n_{o,Na} + \bar{P}_K(\bar{P}_{Na}+J) n_{o,K}\right]} \right] \tag{8.16}$$

results - an expression for the equilibrium membrane potential written in terms of experimentally obtainable parameters.

To obtain an expression for $n_{i,Na}$ substitute (8.16) into (8.13) to give

$$n_{i,Na} = \frac{\bar{P}_K \bar{P}_{Na} n_{o,Na}(n_{o,Na} + n_{o,K} + n_{o,Cl})}{2\left[\bar{P}_{Na}(\bar{P}_K+J/r) n_{o,Na} + \bar{P}_K(\bar{P}_{Na}+J) n_{o,K}\right]} \tag{8.17}$$

In a similar fashion,

$$n_{i,K} = \left[(n_{o,Na} + n_{o,K} + n_{o,Cl})/2\right] \cdot \left[1 - \frac{\bar{P}_K \bar{P}_{Na} n_{o,Na}}{\bar{P}_{Na}(\bar{P}_K+J/r) n_{o,Na} + \bar{P}_K(\bar{P}_{Na}+J) n_{o,K}} \right] \tag{8.18}$$

and

$$n_{i,Cl} = \frac{2n_{o,Cl}\left[\bar{P}_{Na}(\bar{P}_K+J/r) n_{o,Na} + \bar{P}_K(\bar{P}_{Na}+J) n_{o,K}\right]}{\bar{P}_K(\bar{P}_{Na}+J)(n_{o,Na} + n_{o,K} + n_{o,Cl})} \tag{8.19}$$

An equation for the final unknown, cellular volume, may be obtained by substituting equations 8.17 through 8.19 into (9.7).

It is instructive to examine the effects of very low ($J \simeq 0$) and very high ($J \gg 0$) pumping rates on the internal ionic concentrations, membrane potential, and cell volume. Notationally it will be less complex to define

$$\Gamma_1 = n_{o,Na} + n_{o,K} + n_{o,Cl} \tag{8.20}$$

and

$$\Gamma_2 = \bar{P}_{Na}(\bar{P}_K+J/r)n_{o,Na} + \bar{P}_K(\bar{P}_{Na}+J)n_{o,K} \tag{8.21}$$

With these definitions,

$$n_{i,Na} = \bar{P}_K\bar{P}_{Na}n_{o,Na}\Gamma_1/2\Gamma_2 \tag{8.22a}$$

$$n_{i,K} = (\Gamma_1/2) - n_{i,Na} \tag{8.22b}$$

$$n_{i,Cl} = 2n_{o,Cl}\Gamma_2/\bar{P}_K(\bar{P}_{Na}+J)\Gamma_1 \tag{8.22c}$$

$$\varphi_m = -\ln\left[\bar{P}_K(\bar{P}_{Na}+J)\Gamma_1/2\Gamma_2\right] \tag{8.22d}$$

$$V = 2Z_A A_T \Gamma_1 P_K(P_{Na}+J)/\left[\Gamma_1^2 P_K(P_{Na}+J)-4n_{o,Cl}\Gamma_2\right] \tag{8.22e}$$

For low pumping rates ($J \simeq 0$), and an ECF containing no impermeable ions (i.e., $n_{o,K} + n_{o,Na} = n_{o,Cl}$),

$$\begin{aligned} n_{i,Na} &\rightarrow n_{o,Na} \\ n_{i,K} &\rightarrow n_{o,K} \\ n_{i,Cl} &\rightarrow n_{o,Cl} \\ \varphi_m &\rightarrow 0 \\ V &\rightarrow \infty \end{aligned} \tag{8.23}$$

Thus as expected, in the absence of active transport the intracellular ionic concentrations approach those of the ECF, the membrane potential falls to zero and cell volume increases enormously.

For large pumping rates ($J \gg 0$) it is easy to show that

$$\begin{aligned} n_{i,Na} &= \frac{\bar{P}_K\bar{P}_{Na}n_{o,Na}\Gamma_1}{2J(\bar{P}_{Na}n_{o,Na} + \bar{P}_K n_{o,K})} \\ n_{i,K} &= \Gamma_1/2 \\ n_{i,Cl} &= \frac{2n_{o,Na}(\bar{P}_{Na}n_{o,Na} + r\bar{P}_K n_{o,K})}{r\bar{P}_K\Gamma_1} \\ \varphi_m &= -\ln\left[\frac{r\bar{P}_K\Gamma_1}{2(\bar{P}_{Na}n_{o,Na} + r\bar{P}_K n_{o,K})}\right] \end{aligned} \tag{8.24}$$

and

$$V = \frac{2A_T r\bar{P}_K\Gamma_1}{r\bar{P}_K\Gamma_1^2 - 4n_{o,Cl}(\bar{P}_{Na}n_{o,Na} + r\bar{P}_K n_{o,K})}$$

The results noted above for the effects of stopping active transport are reversed when it is speeded up. Thus for $(\bar{P}_{Na}/r\bar{P}_K) < 1$, $n_{i,K}$ increases and $n_{i,Na}$ and $n_{i,Cl}$ decrease, φ_m becomes more negative and the cell volume decreases.

How well does this *ad hoc* formulation predict actual intracellular ionic concentrations and the resting membrane potential? To check on this some data is needed and I take as representative the data of Table 8.1 appropriate for a squid giant axon immersed in sea water.

TABLE 8.1

Parameter	Value	Reference or Remark
$n_{o,Na}$	460 mM/L	Normal for sea water
$n_{o,K}$	10.4	Normal for sea water
$n_{o,Cl}$	470	Insures extracellular electro-neutrality
P_K	1.8×10^{-6} cm/sec	Hodgkin and Katz (1949)
$P_K:P_{Na}$	1:0.04	Hodgkin and Katz (1949)
P_{Na}	0.07×10^{-6} cm/sec	From above
$\bar{P}_K$	0.43×10^{-6} cm/sec	$\bar{P}_K = P_K \varphi_m [1-\exp(-\varphi_m)]^{-1}$
$\bar{P}_{Na}$	0.02×10^{-6} cm/sec	with constant field assumption and $T = 10^{\circ}C$
J	0.8×10^{-6} cm/sec	Hodgkin and Keynes (1955)
r	2	Hodgkin and Keynes (1955)
Γ_1	940 mM/L	Equation 8.20
Γ_2	2.24×10^{-14} M-cm^2/L-sec^2	Equation 8.21

Table 8.1. Measured and derived squid giant axon parameters related to active transport.

From these data and equations 8.22a-d, I calculate that $n_{i,Na} = 83$ mM/L, $n_{i,K} = 387$ while $n_{i,Cl} = 57$. The resting potential is calculated at -53 mV. All of these figures are well within those found experimentally for a squid giant axon, and we must conclude that this formulation describing the equilibrium situation across the squid giant axon membrane is reasonably complete.

C. Active Transport and the Recovery from Excitation

In Chapter 4 I extensively discussed the movement of Na^+ and K^+ across the plasma membrane during excitation. In squid there is an inward transfer of sodium, Δ = 4 pM/cm^2-impulse, and a corresponding efflux of K^+. While it is true that the effect of a few action potentials on the intracellular concentrations of Na^+ and K^+ will be negligible, the transmembrane movement of Na^+ and K^+ associated many action potentials will tend to increase intracellular Na^+ and decrease intracellular K^+.

It is possible to quantitatively examine this process, and this section is devoted to that end. I assume that action potentials are occurring at a frequency F (sec^{-1}) in a fibre of volume V, plasma membrane area A, and diameter d. To a first approximation, the equations connecting the intracellular ionic concentrations with their net fluxes will be of the form

$$V(dn_i/dt) = -AM_i^{net}$$

wherein $M_i^{net} = J_i^{net}/e$ and j_i^{net} for Na^+, K^+ and Cl^- are given by equations 8.5. In writing this equation I have assumed that the rate of change of cell volume is negligible with respect to the rate of concentration changes. Thus for Na^+, K^+ and Cl^-:

$$\frac{dn_{i,Na}}{dt} + \frac{4}{d}(\bar{P}_{Na}+J)n_{i,Na} = \frac{4}{d}\bar{P}_{Na}n_{o,Na}\alpha \tag{8.25a}$$

$$\frac{dn_{i,K}}{dt} + \frac{4}{d}\bar{P}_K n_{i,K} - \frac{4J}{dr}n_{i,Na} = \frac{4}{d}\bar{P}_K n_{o,K}\alpha \tag{8.25b}$$

and

$$\frac{dn_{i,Cl}}{dt} + \frac{4\bar{P}_{Cl}}{d}n_{i,Cl} = \frac{4\bar{P}_{Cl}}{d}n_{o,Cl}\alpha^{-1} \tag{8.25c}$$

From equation 8.14, (8.25b) can be transformed into a second equation involving $n_{i,Na}$ and α; viz.

$$\frac{dn_{i,Na}}{dt} + \frac{4}{d}(\bar{P}_K+\frac{J}{r})n_{i,Na} = \frac{2\bar{P}_K\Gamma_1}{d} - \frac{4}{d}P_K n_{o,K}\alpha \tag{8.26}$$

Eliminating α between (8.25a) and (8.26) gives a first order differentail equation in $n_{i,Na}$ alone,

$$\frac{dn_{i,Na}}{dt} + \beta n_{i,Na} = \gamma \tag{8.27}$$

wherein

$$\beta = \frac{4\Gamma_2}{d(\bar{P}_{Na}n_{o,Na} + \bar{P}_K n_{o,K})} \tag{8.28}$$

and

$$\gamma = \frac{2\Gamma_1 \bar{P}_{Na}\bar{P}_K n_{o,Na}}{d(\bar{P}_{Na}n_{o,Na} + \bar{P}_K n_{o,K})} \tag{8.29}$$

Rewriting (8.27) with the aid of the integrating factor $\exp(\beta t)$ and integrating between $t = (n-1)/F$ (the time at which the n-1 action potential occurred) and t gives

$$\int_{\left[n_{i,Na}\left(\frac{n-1}{F}\right)+\frac{4\Delta}{d}\right]\exp\left[\beta(n-1)/F\right]}^{n_{i,Na}(t)\exp(\beta t)} d\left[n_{i,Na}(x)\exp(\beta x)\right] = \gamma \int_{(n-1)/F}^{t} \exp(\beta x)\,dx$$

or

$$n_{i,Na}(t) = \left[n_{i,Na}\left(\frac{n-1}{F}\right) + (4\Delta/d)\right]\exp\left[\beta(n-1)/F\right]\exp(-\beta t) + (\gamma/\beta)\left[1-\exp\left[\beta(n-1)/F\right]\exp(-\beta t)\right] \tag{8.30}$$

To determine $n_{i,Na}\left[(n-1)/F\right]$ set $t = n/F$ so equation 8.30 becomes a linear first order difference equation of the form

$$x(n) = ax(n-1) + b$$

which has the general solution

$$x(n) = x(0)a^n + b(1-a^n)/(1-a)$$

Thus, identifying $a = \exp(-\beta/F)$, $b = (4a\Delta/d)+(\gamma/\beta)(1-a)$, $x(0) = n_{i,Na}(0)$ (the internal sodium concentration before stimulation commenced) as given by (8.22a) and realizing that $(\gamma/\beta) = n_{i,Na}(0)$, the solution (8.30) becomes

$$n_{i,Na}\left(\frac{n-1}{F} < t \leq \frac{n}{F}\right) = n_{i,Na}(0) + \frac{4\Delta}{d}\,\frac{1-\exp(-\beta n/F)}{1-\exp(-\beta/F)}\exp\left[\beta(n-1)/F\right]\exp(-\beta t) \tag{8.31}$$

Given this expression for the intracellular sodium concentration after n action potentials occurring at a frequency F, and equation 8.14, an expression for $n_{i,K}(t)$ is immediately available:

$$n_{i,K}\left(\frac{n-1}{F} < t \leq \frac{n}{F}\right) = n_{i,K}(0) - \frac{4\Delta}{d}\,\frac{1-\exp(-\beta n/F)}{1-\exp(-\beta/F)}\exp\left[\beta(n-1)/F\right]\exp(-\beta t) \tag{8.32}$$

Thus, as expected, $n_{i,Na}$ increases and $n_{i,K}$ decreases as a consequence of excitation. The extent of the changes occurring are dependent on cell diameter (d) and the characteristic rate constant for this process, β, allthough βd is diameter independent.

Combining equations 8.25a and 8.31 gives an expression for α of the form

$$\alpha\left(\frac{n-1}{F} < t \leq \frac{n}{F}\right) = \alpha(0) + \frac{4\Delta}{d} \frac{\bar{P}_{Na} - \bar{P}_K + J(1-r^{-1})}{\bar{P}_{Na}, n_{o,Na} + \bar{P}_K n_{o,K}} \times \tag{8.33}$$

$$\frac{1-\exp(-\beta n/F)}{1-\exp(-\beta/F)} \exp\left[\beta(n-1)/F\right] \exp(-\beta t)$$

Limiting expressions for $n_{i,Na}$, $n_{i,K}$ and α after many action potentials ($n \to \infty$) are easily obtained and are

$$n_{i,Na}(\infty) = n_{i,Na}(0) + \frac{4\Delta}{d} \frac{\exp(-\beta/F)}{1-\exp(-\beta/F)} \tag{8.34}$$

$$n_{i,K}(\infty) = n_{i,K}(0) - \frac{4\Delta}{d} \frac{\exp(-\beta/F)}{1-\exp(-\beta/F)} \tag{8.35}$$

and

$$\alpha(\infty) = \alpha(0) + \frac{4\Delta}{d} \frac{\bar{P}_{Na} - \bar{P}_K + J(1-r^{-1})}{\bar{P}_{Na} n_{o,Na} + \bar{P}_K n_{o,K}} \frac{\exp(-\beta/F)}{1-\exp(-\beta/F)} \tag{8.36}$$

respectively.

From the data of Table 8.1 values of the parameter β may be calculated for a variety of fibre diameters. In Table 8.2 I list four such values and it is obvious that, except for exceptionally low stimulation rates in very small fibres ($F \ll \beta$), the term

TABLE 8.2

d(μ)	β(sec^{-1})
1	6.6×10^{-2}
10	6.6×10^{-3}
100	6.6×10^{-4}
500	3.3×10^{-4}

Table 8.2. Calculated values of β, from the data of Table 8.1 and equation 8.28, as a function of cell diameter d.

$\exp(-\beta/F)/\left[1-\exp(-\beta/F)\right]$ appearing in equations 8.34-8.36 is well approximated by (F/β). Thus

$$n_{i,Na}(\infty) \cong n_{i,Na}(0) + (4\Delta F/\beta d) \tag{8.37a}$$

$$n_{i,K}(\infty) \cong n_{i,K}(0) - (4\Delta F/\beta d) \tag{8.37b}$$

and

$$\alpha(\infty) \cong \alpha(0) + \Delta F\left[\bar{P}_{Na} - \bar{P}_K + J(1-r^{-1})\right]/\Gamma_2 \tag{8.37c}$$

to predict maximal changes in intracellular sodium and potassium concentrations of $\pm$ 2.42 F mM/L (F in sec^{-1}) respectively. Note also from Table 8.2 that the factor βd has a value of 6.6×10^{-6} cm/sec and is independent of cell diameter, thus implying that these predicted concentration changes are not a function of cell size.

Another point of interest is raised by (8.37c) or, for that matter, by (8.33) or (8.36). From these equations it is obvious that stimulation may affect the membrane potential in quite diverse ways. For an electroneutral pump (r=1) the effect of stimulation will always be a depolarization of the cell membrane since $\bar{P}_K > \bar{P}_{Na}$. However, for an electrogenic pump with $r > 1$, $\bar{P}_{Na} + J(1-r^{-1}) < \bar{P}_K$ implies a depolarizing response to sustained activity while $\bar{P}_{Na} + J(1-r^{-1}) > \bar{P}_K$ will give a hyperpolarization.

PROBLEMS

8.1. The intracellular pH of a muscle cell is 7.0 and the pH of the extracellular fluid is 7.4. What is the equilibrium potential for H^+? The membrane potential is -90 mV. Is H^+ in equilibrium across the membrane? If it is or is not in equilibrium, what are the implications of that finding?

8.2. In the Hodgkin-Katz equation (7.18) for the resting potential of a cell (whose membrane is permeable only to Na^+, K^+ and Cl^-) there is apparently no explicit accounting made of the fact that Na^+ and K^+ are actively transported against their respective electrochemical gradients. Was there a mistake in the derivation of this relationship, or is the existence of active transport implicit? Explain.

8.3. Suppose it is postulated that membrane active transport can be described by a constant (with respect to time and distance through the membrane) force F per ion actively transported. Thus for ions actively transported an additional ionic current density, $q\mu nF$, must be added to the current density, due to the "passive" influences of diffusion ($-qDdn/dx$) and electric field within the membrane ($q^2 n\mu E$). Thus,

$$j = -qD(dn/dx) + q^2\mu nE + q\mu nF \qquad (*)$$

a. Derive an expression similar to the Nernst equation that gives the relation between the membrane potential difference, φ_m, the ionic concentrations on each side of the membrane (n_i and n_o) and F when j = 0. State all of your assumptions clearly.

b. Assuming that E is constant within the membrane, integrate equation * to give an expression for j as a function of φ_m, n_i, n_o and F. Compare the behavior of your expression with the Goldman equation (7.8) predicted behavior.

8.4. A giant axon from the sabled earthworm has a membrane primarily permeable to Na^+, K^+, and Cl^- at rest. Also, tracer studies indicate that there is active transport of Na^+ and K^+ across the membrane that has identical characteristics to the Na^+-K^+ active transport system of the squid giant axon. The normal pumping rate, J, is 5×10^{-7} cm/sec; the ratio $\bar{P}_{Na}/\bar{P}_K = 1.5 \times 10^{-3}$; and normally $r = 3$ (3 Na^+ are actively transported for 1 K^+). Take $\bar{P}_K = 10^{-6}$ cm/sec.

a. Calculate and plot $n_{i,Na}$, $n_{i,K}$, $n_{i,Cl}$ and φ_m as a function of:

   i. $n_{o,Na}$ for the range 0 to 500 mM/L with $n_{o,K} = 20$, $n_{o,Cl} = 500$.

   ii. $n_{o,K}$ for the range 0 to 500 mM/L with $n_{o,Na} = 400$, $n_{o,Cl} = 500$.

b. The addition of compound GLUP to the solution bathing the axon alters $\bar{P}_{Na}/\bar{P}_K$ so it is 0.02. Repeat section (a) for this case.

c. Substance GORP decreases r so that it is now one (1). Repeat section (a) for this case.

8.5. In a recent experiment the axoplasm of a squid giant axon was extruded, thus eliminating the impermeant anions, and replaced with a solution containing Na^+, K^+, and Cl^- as the only ions. The extracellular fluid also contained Na^+, K^+, and Cl^- and it was found that active transport of Na^+ and K^+ was still taking place in a normal fashion.

Assuming:

a. The characteristics of Na^+-K^+ active transport to be normal;

b. The passive transport of Na^+, K^+, and Cl^- to be governed by electrodiffusion theory;

c. The intracellular fluid be electrically neutral;

d. That the extracellular concentration of Na^+, K^+, and Cl^- are experimentally maintained.

a. Derive expressions for $n_{i,Na}$, $n_{i,K}$, $n_{i,Cl}$ and the membrane potential, φ_m.

b. What is φ_m when active transport is halted ($J = 0$)? What are the internal ionic concentrations?

c. Generally, is there any difference in osmotic pressure across the membrane? Prove your assertation.

CHAPTER 9. ADMITTANCE PROPERTIES OF THE ELECTRODIFFUSION EQUATIONS.

Early attempts to characterize the electrical properties of excitable membranes utilized alternating current impedance measurements on the giant axon of *Loligo* and the excitable cell from *Nitella*. Transverse impedance measurements on both of these systems in the resting state gave data that was interpreted as arising from a membrane capacity (C_m) of approximately 1μF/cm^2 in parallel with a membrane resistance (R_m) on the order of 10^3 ohm-cm^2 (Curtis and Cole 1938; Cole and Hodgkin, 1939).

Further experiments, utilizing transverse impedance measurements during excitation in both *Nitella* and squid, conclusively demonstrated that the major change during an action potential was a transient decrease in R_m (Cole and Curtis, 1938, 1939). In addition to this transient change in membrane resistance during excitation, it was noted that there was a small but consistent change in the reactive component of the membrane impedance. The effect could always be interpreted as a decrease in membrane capacity, and was most pronounced during the falling phase of the action potential. Similar results were obtained during the passage of current through the membrane (Cole and Baker, 1941).

Cole and Baker (1941), using longitudinal impedance measurements, were able to extend their *Loligo* axon membrane data to frequencies much lower than was possible with transverse measurements. Some interesting phenomena came to light at frequencies below 250 cps. In this frequency range there was a clear indication that the reactive components of the membrane impedance could, under some conditions, become inductive. The data indicated the existence of a membrane inductive element on the order of 0.2 H-cm^2. Apparently, the impedance properties of the squid giant axon membrane derived from two different components. One, a high frequency region, always exhibited a capacitive reactance; while a second, low frequency, portion had a reactive element that was either capacitive or inductive. A number of experiments indicated that the low frequency reactive portion was likely to become inductive in lowered external calcium solutions or in high external potassium solution. Cole (1941,1949) hypothesized that the low frequency characteristics of the membrane impedance were associated with potassium movement across the excitable membrane. Oscillatory behaviour, also indicative of the existence of a membrane inductive element, was noted by Hodgkin and Rushton (1946) in a small signal time domain analysis on the giant axon of _Homarus_. Similar results were noted by Weidmann (1950) in _Nitella_.

With the advent of the characterization and detailed analysis of excitable membrane behaviour in the time domain (Hodgkin and Huxley, 1952) it became possible to understand the origin of the frequency dependent behaviour noted earlier. A small signal analysis of the Hodgkin-Huxley equations about the resting potential leads to a frequency domain equivalent circuit representation for the excitable membrane (Chandler, Fitzhugh, and Cole, 1962; Cole, 1968). This analysis indicates that at low frequencies the reactance of the membrane should be inductive, with the majority of the contribution coming from potassium conductance activation. A small amount of the inductance derives from sodium conductance inactivation. The reactive element of the sodium conductance activation is capacitive, so at higher frequencies the sodium channel exhibits a net capacity.

Mauro, Conti, Dodge and Schor (1970) carried out a detailed experimental study of the small signal time domain electrical responses of the membrane of the giant axon of _Todarodes sagittatus_. They found excellent agreement between their results and the behaviour predicted from a small signal analysis of the Hodgkin-Huxley equations.

These small signal results from the linearized Hodgkin-Huxley equations allow a qualitative understanding of the phenomena mentioned earlier. It was noted that during excitation the membrane impedance does not behave as if there was a pure resistive decrease, but rather as if there is a change in reactance and that the deviation from the ideal is maximal during the falling phase of the action potential. This may be accounted for by the inductive behaviour of the potassium activation system and the capacitive behaviour of the sodium activation system respectively. Both of these effects would tend to increase the series reactance of the membrane, interpreted as a decrease in membrane capacitance.

It seems likely that the effects of elevated external potassium on membrane impedance are due to a decrease in membrane potential. The primary effect of this depolarization is to activate the potassium conductance, and thus to increase the apparent membrane inductance. The fact that the inductive effect decreases past a certain potassium concentration may be due to competitive effects of the sodium system. The effects of external calcium alterations admit to a similar explanation. Frankenhaeuser and Hodgkin (1957) have shown that an increase in calcium affects the membrane much like a hyperpolarization, which would tend to remove the potassium system from activity and hence the attendant inductive effect. The effects of low calcium concentrations are much like the effects of high external potassium, and can be explained in the same fashion.

Although use of the linearized Hodgkin and Huxley equations gives qualitative insight into the ionic origin of the impedance properties noted in excitable systems; it gives little, if any, feeling for the mechanisms associated with ion transport that might give rise to inductive or capacitive effects.

In this chapter I analyse the small-signal admittance properties of a membrane model system in which it is assumed that ions transversing the membrane obey the Nernst-Planck electrodiffusion equation. Thus, the effects of ionic diffusion coefficients, equilibrium potentials, and the membrane potential in determining the reactive and resistive behaviour of membrane transport elements may be evaluated.

The decision to base an analysis of this type on electrodiffusion theory was made on a two-fold basis. Cole (1968) has extensively reviewed the properties of excitable membrane systems, and the attempts to model them. Although electrodiffusion theory, in the Nernst-Planck formulation, fails to explain or predict behaviour contained in the full Hodgkin-Huxley formulation of axonal properties it does offer considerable insight into sub-threshold properties. Cohen and Cooley (1965) and Cooley, Dodge and Cohen (1965) observed that electrodiffusion model systems display time domain computed responses suggestive of systems with mixed capacitive-inductive reactance properties. These considerations, coupled with Sandbloms (1972) computed frequency domain behaviour in a linearized Nernst-Planck system makes a general analytical study of anomalous reactance properties in electrodiffusion systems potentially interesting. An analogous analysis in the time domain was carried out by Cole (1965).

A. ANALYSIS

I assume that there exist discrete, spatially localized, regions within the membrane through which ions move (Hille, 1970). In this analysis the admittance properties of one class of ions moving through one type of ion permeable membrane region will be examined. The dimensionless ionic current density, I (amp/cm^2) is given by equation 6.18:

$$I = -Z\left[\frac{\partial N}{\partial X} + ZN\frac{\partial \varphi}{\partial X}\right] \tag{6.18}$$

In addition to equation 6.18 the continuity equation is required:

$$\frac{\partial I}{\partial X} = -Z\frac{\partial N}{\partial T} \tag{6.12}$$

To deal with equations 6.12 and 6.18 a knowledge of $\varphi(X,T)$ is needed; this is normally derived from Poisson's equation, 6.17. However, in accord with much other work on electrodiffusion models, attention is directed here to situations satisfying the Goldman (1943) constant field assumption.

Determination of the Membrane Admittance

The behaviour of the electrodiffusion model for small perturbing time dependent voltages, $\delta\varphi \exp(-jWT)$ with $j = \sqrt{-1}$, about a steady state voltage φ_o is of primary interest so take

$$I(X,T) = I_o(X) + \delta i \exp(-jWT)$$

$$\varphi(X,T) = \varphi_o(X) + \delta\varphi \exp(-jWT)$$

$$N(X,T) = N_o(X) + \delta n \exp(-jWT)$$

When these relations are substituted into (6.18) and (6.12), two sets of equations are obtained. The first,

$$I_o = -Z\left[\frac{dN_o}{dX} + ZN_o \frac{d\varphi_o}{dX}\right] \qquad (9.1)$$

describes the steady state ion flow; while the second set,

$$\delta i = -Z\left[\frac{d\delta n}{dX} + ZN_o \frac{d\delta\varphi}{dX} + Z\delta n \frac{d\varphi_o}{dX}\right] \qquad (9.2)$$

and

$$\frac{d\delta i}{dX} = jWZ\delta n \qquad (9.3)$$

describe the perturbed state.

<u>Steady state</u>. Utilizing the Goldman constant field assumption and referring the potential at the inner border of the membrane ($\varphi = \varphi_i$ at $X = 0$) to the potential at the outer membrane border ($\varphi = \varphi_d$ at $X = 1$), $\varphi_o(X) = \varphi_m(1-X)$, where $\varphi_m = \varphi_i - \varphi_d$. Thus $(d\varphi_o/dX) = -\varphi_m$ and (9.1) becomes

$$I_o = -Z\left[\frac{dN_o}{dX} - ZN_o\varphi_m\right] \qquad (9.4)$$

Given the boundary conditions $N_o = 1(N_d)$ at $X = 0(1)$, assumed to be experimentally maintained, the integration of (9.4) yields

$$N_o = \alpha\exp(Z\varphi_m X) + \beta \qquad (9.5)$$

wherein

$$\alpha = (1 - N_d)/(1 - \exp Z\varphi_m) \qquad (9.6)$$

and

$$\beta = (N_d - \exp Z\varphi_m)/(1 - \exp Z\varphi_m) \qquad (9.7)$$

$\varphi_e = Z^{-1}\ln N_d$ is the ionic equilibrium potential.

<u>Perturbed state</u>. Again make the constant field assumption with respect to the perturbing voltage, $\delta\varphi$, so $\delta\varphi = \delta\varphi_m(1-X)$ and (9.2) becomes

$$\delta i = -Z\left[\frac{d\delta n}{dX} - ZN_o\delta\varphi_m - Z\varphi_m\delta n\right] \qquad (9.8)$$

or, in conjunction with (9.3),

$$-jW\delta n = \frac{d^2\delta n}{dX^2} - Z\,\delta\varphi_m \frac{dN_o}{dX} - Z\varphi_m \frac{d\delta n}{dX} \tag{9.9}$$

If the spatial Fourier transform of δn is denoted by $\tilde{n}(s,W)$ and (9.5) is used to calculate (dN_o/dX), equation 9.9 may be solved for $\tilde{n}$ to give

$$\tilde{n}(s,W) = \frac{Z^2\alpha\varphi_m\,\delta\varphi_m}{(s - Z\varphi_m)[s(s - Z\varphi_m) + jW]} \tag{9.10}$$

Further denoting the Fourier transform of δi by $\tilde{i}(s,W)$

$$\tilde{i}(s,W) = Z^2\left[\frac{\beta\delta\varphi_m}{s} + \frac{\alpha\delta\varphi_m}{s-Z\varphi_m}\right] - Z(s - Z\varphi_m)\tilde{n}(s,W) \tag{9.11}$$

results from (9.8).

If equation 9.10 for $\tilde{n}$ is substituted into (9.11) for $\tilde{i}$, the (spatial) Fourier transformed admittance, $\tilde{Y}(s,W) = \tilde{i}/\delta\varphi_m$ becomes

$$\tilde{Y}(s,W) = Z^2\left[\frac{\beta}{s} + \frac{\alpha}{s - Z\varphi_m} - \frac{\alpha Z\varphi_m}{s(s-Z\varphi_m) + jW}\right] \tag{9.12}$$

from equation 9.11.

Let the real and imaginary parts of $\tilde{Y}(s,W)$ be given by $\tilde{G}(s,W)$ and $\tilde{B}(s,W)$ respectively, so $\tilde{Y}(s,W) = \tilde{G}(s,W) + j\tilde{B}(s,W)$. Thus, from (9.12),

$$\tilde{G}(s,W) = Z^2\left[\frac{\beta}{s} + \frac{\alpha}{s - Z\varphi_m} - \frac{\alpha Z\varphi_m s(s-Z\varphi_m)}{s^2(s-\varphi_m Z)^2 + W^2}\right] \tag{9.13}$$

and

$$\tilde{B}(s,W) = \frac{\alpha Z^3\varphi_m W}{s^2(s-Z\varphi_m)^2 + W^2} \tag{9.14}$$

It is possible from (9.12), or from (9.13) and (9.14), to write an expression for $Y(X,W)$, or $G(X,W)$ and $B(X,W)$. However, the resulting expressions are so complicated that the essential features I wish to discuss are obscured.

B. THE BEHAVIOUR OF THE TRANSFORMED MEMBRANE ADMITTANCE

Following Cole and Cole (1941) it will be convenient to discuss the behaviour of $Y(s,W)$ in the complex admittance plane, displaying $B(s,W)$ as a function of $G(s,W)$ with W as a parametric variable.

From equation 9.13 it is easy to show that the transformed zero and infinite frequency conductances are given by

$$\tilde{G}_o(s) = \lim_{W\to 0}\tilde{G}(s,W) = \frac{Z^2}{s} \tag{9.15}$$

and

$$G_{\infty}(s) = \lim_{W \to \infty} \tilde{G}(s,W) = Z^2\left[\frac{\beta}{s} + \frac{\alpha}{s - Z\varphi_m}\right] \quad (9.16)$$

respectively. Let

$$\tilde{G}_1(s) = \left[\tilde{G}_{\infty}(s) - \tilde{G}_o(s)\right]/2 \quad (9.17)$$

and

$$\tilde{G}_2(s) = \left[\tilde{G}_{\infty}(s) + \tilde{G}_o(s)\right]/2 \quad (9.18)$$

so it is a simple matter to show that

$$\left[\tilde{G}(s,W) - \tilde{G}_2(s)\right]^2 + \left[\tilde{B}(s,W)\right]^2 = \left[\tilde{G}_1(s)\right]^2 \quad (9.19)$$

Thus, from equation 9.19, as W ranges from $-\infty$ to $+\infty$, $\tilde{Y}(s,W)$ describes a circle in the complex admittance plane of radius $\tilde{G}_1(s)$ with its centre on the $\tilde{G}(s,W)$ axis at $\tilde{G}_2(s)$. Only non-negative frequencies are of interest, so for this range of W the relationship is a semi-circle. Further, from the requirements for the existence of the transforms in (9.13) and (9.14) and the expressions for α and β given in (9.6) and (9.7)

$$\left.\begin{array}{c} \tilde{G}_{\infty}(s) - \tilde{G}_o(s) \\ \dfrac{\partial \tilde{G}(s,W)}{\partial W} \\ \tilde{B}(s,W) \end{array}\right\} > 0 \ (< 0)$$

if $Z(\varphi_m - \varphi_e) > 0 (< 0)$. From all of the above considerations the complex admittance plane behaviour of $\tilde{Y}(s,W)$ is as illustrated in Figure 9.1.

The total admittance of a circuit consisting of a conductance G_1 in parallel with a series combination of a conductance G_2 and a susceptance B may be written as

$$Y_T = G_1 + \frac{jBG_2}{G_2 + jB}$$

or upon separation into real and imaginary portions

$$\mathrm{Re}\, Y_T = G_1 + \frac{G_2B^2}{G_2^2 + B^2}$$

and

$$\mathrm{Im}\, Y_T = \frac{G_2^2B}{G_2^2 + B^2}$$

The susceptance of a pure capacitance is $B = WC$, while for an inductance $B = -(WL)^{-1}$; therefore, comparison of the circuit equations from Figure 9.2 with equations 9.6 and 9.7 immediately leads to the identification of $\tilde{G}_{\infty}(s)$ with G_1, $\tilde{G}_{\infty}(s) - \tilde{G}_o(s)$ with G_2, and B with a capacitor $C = \alpha Z^3\varphi_m/W^2$ when $\tilde{B}(s,W)$ is positive and with an inductor

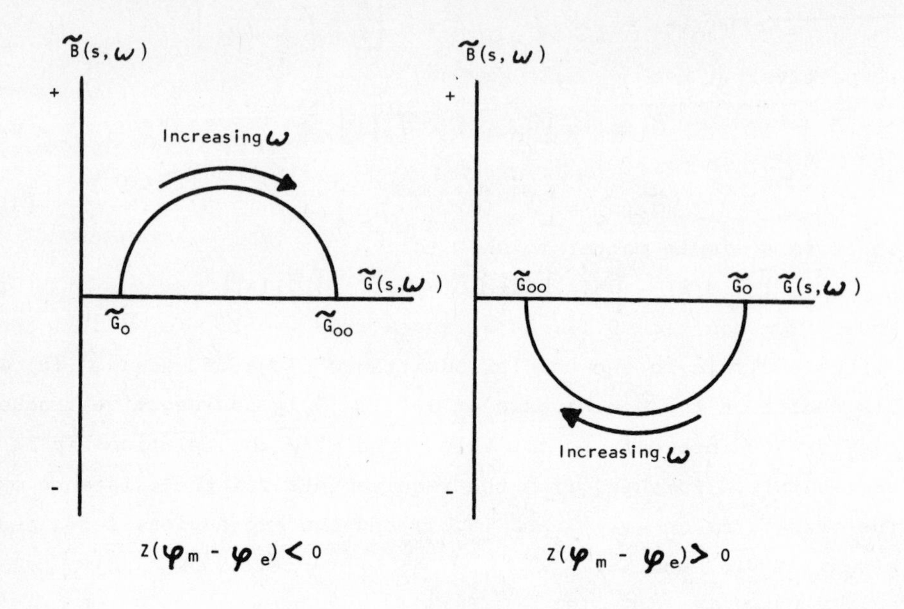

Figure9.1. The behaviour of the (spatially) Fourier transformed admittance $\tilde{Y}(s,W) = \tilde{G}(s,W) + j\tilde{B}(s,W)$, obtained from a small signal analysis of the Nernst-Planck single ion equation and the continuity equation. Susceptance, $\tilde{B}(s,W)$, is shown as a function of conductance $\tilde{G}(s,W)$, with W as a parametric variable. $\tilde{G}_o$ and $\tilde{G}_\infty$ are, respectively, the zero and infinite frequency conductances and are given by equations 9.15 and 9.16. A positive susceptance corresponds to a capacitance, negative to an inductance.

The left hand plot shows the admittance plane behaviour for $Z(\varphi_m - \varphi_e) < 0$ where Z is ionic valence, φ_m and φ_e are the membrane and ionic equilibrium potentials respectively. On the right is the same diagram for $Z(\varphi_m - \varphi_e) > 0$.

$L = -(\alpha Z^3 \varphi_m)^{-1} > 0$ when $\tilde{B}(s,W) < 0$. Thus, the admittance plane properties of Figure 9.1 may be completely derived from the equivalent electrical circuit representation shown in Figure 9.2.

Therefore, any ion transport system described by the Nernst-Planck electrodiffusion equations with constant electric field will display "anomalous reactance" properties if there is an ionic concentration gradient across the membrane and the membrane potential (φ_m) is not identical to the ionic equilibrium potential (φ_e). In a

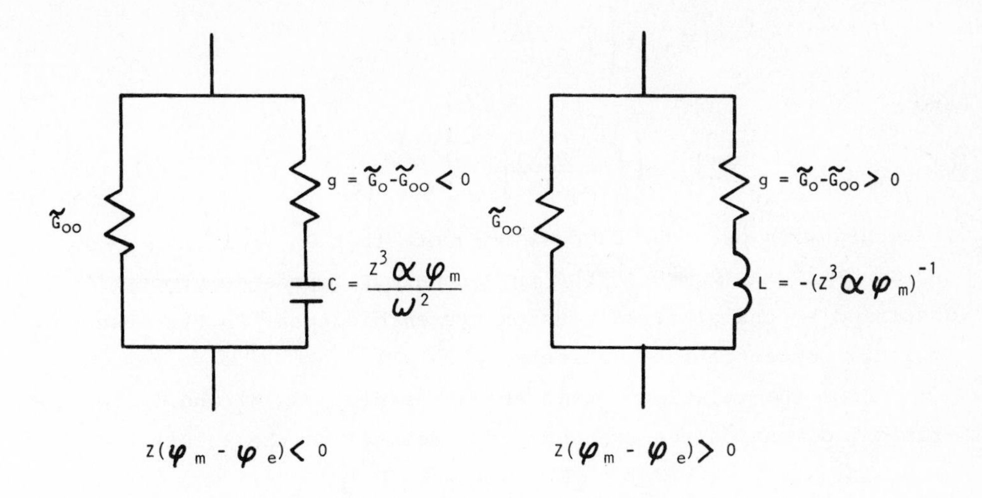

Figure 9.2.The equivalent circuit representations for electro-diffusion membrane elements giving rise to the complex admittance plane properties shown in Figure 9.1.

tissue such as the squid giant axon, with normal intra- and extra-cellular environments, the analysis of this simple electrodiffusion model for ion transport predicts that near the resting potential potassium ion movement should be associated with complex admittance properties containing an anomalous inductive element while sodium movement would be associated with an anomalous capacitive reactance. These results are qualitatively consistent with the experimental behaviour noted earlier, as well as the theoretical predictions of the small signal frequency domain approximation of the Hodgkin-Huxley equations.

To this point the spatial Fourier transforms $\tilde{G}_o(s)$ and $\tilde{G}_\infty(s)$ of $G_o(X)$ and $G_\infty(X)$ have been used exclusively. In a quantitative comparison between the Nernst-Planck formulations presented here and the only set of data available, the Hodgkin-Huxley equations in the small signal approximation, it will be essential to have expressions for the integrated zero and infinite frequency conductances. From equations 9.15 and 9.16, $G_o(X)$ and $G_\infty(X)$ are easily integrated to give

$$G_{om} = \left[\int_0^1 \frac{dX}{G_o(X)} \right]^{-1} = Z^2 \tag{9.20}$$

and

$$G_{\infty m} = \left[\int_0^1 \frac{d(X)}{G_\infty(X)} \right]^{-1} = \frac{Z^2 \varphi_{m\beta}}{\varphi_m - \varphi_e} \tag{9.21}$$

In accord with previous discussion, note that $G_{\infty m} \to Z^2$ as $\varphi_m \to \varphi_e$ and $G_{\infty m} \to Z^2$ as $\varphi_e \to 0$. Thus the anomalous impedance properties displayed by the electrodiffusion system disappear in the absence of ionic concentration gradients.

From the relations given above for $B(W)$, $G_o(W)$ and $G_\infty(W)$ the critical dimensionless frequency W_c, defined as the solution of,

$$B(W_c) = \left[G_\infty(W_c) - G_o(W_c) \right] / 2$$

may also be calculated. In comparisons with e.g. the Hodgkin-Huxley parameters, however, W_c is not as interesting as the dimensionless time constant of the system defined by $W_c \tau_c = 1$. When these determinations are carried through

$$\tau_c(\varphi_m, \varphi_e) = \frac{\left[\dfrac{\varphi_{m\beta}(\varphi_m, \varphi_e)}{\varphi_m - \varphi_e} - 1 \right]}{2Z\varphi_{m\alpha}(\varphi_m, \varphi_e)}$$

results. It is easy to show that τ_c has a single maximum at $\varphi_m = \varphi_e$ and $\tau_c \to 0$ for $\varphi_m \to \pm\infty$. Qualitatively these results are as found for the τ_n of the Hodgkin-Huxley formalism describing potassium current kinetics.

C. QUANTITATIVE COMPARISONS WITH DATA

The Hodgkin-Huxley (1952) equations, in the small signal approximation (Mauro, Conti, Dodge and Schor, 1970), afford an excellent opportunity to compare the present predictions of the Nernst-Planck formulation of electrodiffusion theory with experimental data. It is well known (cf. Cole, 1965) that electrodiffusion theory is unable to match the known sodium conductance characteristics in the squid giant axon membrane. However, the potassium channel steady state conductance is at least semi-quantitatively consistent with the predictions of electrodiffusion theory so I will concentrate my attention on the small signal approximation to the Hodgkin-Huxley expression for potassium current.

In the notation of Hodgkin and Huxley a small signal analysis of j_K yields the membrane equivalent circuit of Figure 9.2b where the circuit components are given by

$$G_K(V) = \bar{g}_K n_\infty \tag{9.22}$$

$$g_K(V) = \frac{4 g_K n_\infty^3 (V-V_K)\left\{\frac{d}{dV}\left[\alpha_n(1-n)\right] - n_\infty \frac{d\beta_n}{dV}\right\}}{\alpha_n + \beta_n} \tag{9.23}$$

and

$$L_K(V) = \frac{1}{(\alpha_n + \beta_n) g_K(V)} \tag{9.24}$$

The corresponding circuit values predicted from this study are

$$\bar{G}_K(V) = \frac{e^2 D_K \bar{N}_{iK}}{dKT} \cdot \frac{V+V_r}{V-V_{eK}} \cdot \frac{(\bar{N}_{dK}/\bar{N}_{iK}) - \exp\left[e(V+V_r)/KT\right]}{1 - \exp\left[e(V+V_r)/KT\right]} \tag{9.25}$$

$$\bar{g}_K(V) = \frac{e^2 D_K \bar{N}_{iK}}{dKT} - \bar{G}_K(V) \tag{9.26}$$

and

$$\bar{L}_K(V) = -\frac{d^3 (KT)^2}{e^3 D_K^2 (\bar{N}_{iK}-\bar{N}_{dK})} \cdot \frac{1 - \exp\left[e(V+V_r)/KT\right]}{(V+V_r)} \tag{9.27}$$

In equations 9.25 to 9.27 the theoretical results have been transformed back to dimensional form and all potentials, φ , are expressed relative to the resting potential, V_r. Thus $V = \varphi_m - V_r$ and $V_e = \varphi_e - V_r$.

For the purposes of computation take T = 6.3°C, and d = 75 Å. Data of Hodgkin and Huxley indicate $V_r \simeq -60$ mV and $V_{eK} \simeq -12$ mV. I assume V_r = -60 mV, and V_{eK} = -11.2 mV, thereby implying that (N_{iK}/N_{dK}) = 20, and thus N_{iK} = 0.4 M/L if N_{dK} = .02 M/L. The constant

$$G_o = e^2 D_K \bar{N}_{iK} / dKT$$

appears in all three equations, (9.25 - 9.27), and if D_K has the dimensions ($cm^2 sec^{-1}$), G_o will have the value

$$G_o = 1.95 \times 10^6 \, D_K \text{ mho/cm}^2$$

No good independent estimate of D_K is available, so it is of interest to calculate the values of D_K which will bring the values for $\bar{G}_K$, g_K, and $\bar{L}_K$ into agreement with the values predicted from the Hodgkin-Huxley equations.

Equations 9.22 through 9.24 give, at V = 0 (i.e. the resting potential), $G_K(0) = 3.67 \times 10^{-4}$ mho/cm^2, $g_K(0) = 8.33 \times 10^{-4}$ mho/cm^2, and $L_K(0) = 6.43$ Henry-cm^2. In order to obtain an exact correspondence between the value for $G_K(0)$ predicted by the Nernst-Planck and Hodgkin-Huxley analyses D_K must be 10.66×10^{-10} cm^2/sec. With this value for D_K, $g_K(0) = 1.71 \times 10^{-3}$ mho/cm^2 and $\bar{L}_K(0) = 1.77$ Henry-cm^2. Thus there is a discrepancy in the $\bar{g}_K$ values by a factor of 2.05 and in the $\bar{L}_K$ values by a factor of 3.63.

In Table 9.1 the values of D_K necessary to bring one of the three quantities being examined into exact agreement with values predicted by the Hodgkin-Huxley equations are listed. It would be possible to do a least squares fit between equations 9.22 through 9.24 and 9.25 through 9.27 at V = 0, simultaneously determining the best values of D_K, V_{eK}, and V_r.

$\bar{D}_K \times 10^{10}$ (cm^2/sec)	$\bar{G}_K(0) \times 10^4$ (mho/cm^2)	$\bar{g}_K(0) \times 10^4$ (mho/cm^2)	$\bar{L}_K(0)$ (H-cm^2)
10.66	3.67	17.10	1.77
5.19	1.79	8.33	7.46
5.59	1.93	8.97	6.43
Hodgkin-Huxley	3.67	8.33	6.43

Table 9.1. Predicted values of $\bar{G}_K$; $\bar{g}_K$, and $\bar{L}_K$ at the resting potential for three different potassium diffusion coefficient values and the corresponding values predicted by the Hodgkin-Huxley equations.

Mauro, Conti, Dodge and Schor (1970, Figure 19a) have plotted $G_K(V)$, $g_K(V)$, and $L_K(V)$ as obtained from the Hodgkin-Huxley equations (see equations 9.22 through 9.24) and the equations derived from this analysis, (9.25) through (9.27) show all of the same qualitative features.

One of the most interesting features of Table 9.1 revolves around the values of D_K necessary to bring the predicted values of

$\bar{G}_K(0)$, $\bar{g}_K(0)$, and $\bar{L}_K(0)$ into agreement with their actual values. Using three different pieces of data the necessary values of D_K range between 5.19×10^{-10} and 10.66×10^{-10} cm^2/sec. The diffusion coefficient for K^+ in free solution at physiological concentrations and the same temperature range is some five orders of magnitude larger than these values thus supporting the conclusions of Cole and Moore (1960) and Cole (1968) that K^+ diffusion with the membrane is much more restricted than in free solution. Cole's (1968) estimate for D_K was 5×10^{-10} cm^2/sec.

CHAPTER 10. THE STEADY STATE AGAIN: APPROXIMATIONS TO INVESTIGATE THE ROLE OF MEMBRANE FIXED CHARGE

As discussed in Chapter 5, biological membranes tend to be more permeable to cations than anions and, further, fine discriminations are made by the membrane between various cations and anions. The high cation permeability of most plasma membranes, coupled with demonstrated net negative fixed charge within the membrane (c.f. Chapter 5), raises the question of whether these two phenomena are linked.

In Chapter 5 I mentioned theoretical attempts to link Hodgkin-Huxley parameter voltage shifts in the face of altered intra- or extracellular changes in pH, ionic strength, or divalent ion concentration to surface charge effects. In this short chapter I briefly examine the role that net volume charge in the membrane, or net surface charge at the membrane-electrolyte interface, might play in modifying traditional electrodiffusion predictions of ionic current versus membrane potential relations.

A. Volume Charge Considerations. I assume that there is a volume charge ρ within the membrane that does not vary with position. Thus the dimensionless form of Poissons equation 6.17 becomes

$$\frac{d\bar{E}}{dX} = \bar{\rho}\,\delta^2/\lambda_D^2 = -\frac{d^2\varphi}{dX^2} \qquad (10.1)$$

Using the same nomenclature as in Chapter 9 for boundary values of φ at X=0 and X=1 (φ_1 and φ_d respectively) equation 10.1 is easily integrated to give

$$\varphi(X) = \frac{\bar{\rho}\,\delta^2}{2\lambda_D^2}(X-X^2) + \varphi_m(1-X) \qquad (10.2)$$

wherein $\varphi_m = \varphi_i - \varphi_d$ and I have taken the potential at the outer membrane border (X=1) to be zero (φ_d=0).

Thus the constant volume charge assumption leads to a quadratic potential profile across the membrane composed of a linear portion, identical to that obtained from the constant field assumption, and a superimposed parabolic portion arising from the fixed charge.

The only way that $\varphi(X)$, as given by (10.2), enters into the calculation of I versus φ_m relations based on electrodiffusion theory is through the evaluation of the integral

$$\int_0^1 \exp\left[Z\,\varphi(X)\right] dX$$

that appears in (7.6). From the mean value theorem, for f(X) and g(X) continuous on (0,1) there exists an $\bar{X}$ in that open interval such that

$$\int_0^1 f(X)g(X)dX = f(\bar{X}) \int_0^1 g(X)dX$$

Thus if I set

$$f(X) = \exp\left[\frac{Z\bar{\rho}\,\delta^2}{2\lambda_D^2}(X-X^2)\right]$$

and

$$g(X) = \exp\left[Z\,\varphi_m(1-X)\right]$$

then I may write

$$\int_0^1 \exp\left[Z\,\varphi(X)\right] dX = \exp\left[\frac{Z\bar{\rho}\,\delta^2}{2\lambda_D^2}\bar{X}(1-\bar{X})\right] \frac{\exp(Z\varphi_m)-1}{Z\varphi_m}$$

and in the presence of net fixed volume charge equation 7.8 becomes (in a dimensionless form)

$$I_{vc} = I_G \exp\left[-\frac{Z\bar{\rho}\,\delta^2}{2\lambda_D^2}\bar{X}(1-\bar{X})\right] \qquad (10.3)$$

where I_G denotes the analogous expression obtained with the constant field assumption.

Now $0 < \bar{X} < 1$ so $1-\bar{X} > 0$. Thus if $Z\bar{\rho} > 0 (< 0)$ then bound charge will reduce (increase) the ionic current and conductance below (above) the Goldman equation values. Net negative fixed charge will reduce (increase) anion (cation) current and positive fixed charge will reduce (increase) cation (anion) current. All of these conclusions are intuitively reasonable.

How significant is this effect likely to be? To answer this question, an estimate of (I_{VC}/I_G) is needed from (10.3). Assuming $X = 0.5$ (which will maximize the effect), $\delta = 100$ A, $\varepsilon = 3$, $T = 300^{\circ}$K, $\rho = 3.2 \times 10^7$ coul/mt^3 and $|Z| = 1$ then for $Z\rho > 0$, $(I_{VC}/I_G) = \exp(-0.37) = 0.69$, while for $Z\rho < 0$, $(I_{VC}/I_G) = 1.45$.

To evaluate

$$\int_0^1 \exp\left[Z\varphi(X)\right] dX$$

exactly, with $\varphi(X)$ given by (10.2), I make use of the fact that

$$\int_0^1 \exp\left[-(aX^2 + 2bX + c)\right] dX$$

$$= \frac{1}{2}\sqrt{\frac{\pi}{a}} \exp\left[\frac{b^2-ac}{a}\right]\left\{\operatorname{erf}\left[\sqrt{a} + \frac{b}{\sqrt{a}}\right] - \operatorname{erf}\left[\frac{b}{\sqrt{a}}\right]\right\} \tag{10.4}$$

In employing (10.4), write

$$Z\varphi(X) = -(aX^2 + 2bX + c) \tag{10.5}$$

wherein

$$\begin{aligned} a &= Z\bar{\rho}\,\delta^2/2\lambda_D^{\,2} \\ b &= -(Z/2)\left[(Z\bar{\rho}\,\delta^2/2\lambda_D^{\,2}) - \varphi_m\right] \\ c &= -Z\varphi_m \end{aligned} \tag{10.6}$$

and note that $b = -(a+c)/2$, to give

$$\int_0^1 \exp\left[Z\varphi(X)\right] dX = \sqrt{\frac{\pi}{4a}} \exp\left[(a-c)^2/a\right] \times$$
$$\left\{\operatorname{erf}\left[(a-c)/2\sqrt{a}\right] + \operatorname{erf}\left[(a+c)/2\sqrt{a}\right]\right\} \tag{10.7}$$

Now $\exp Z\varphi(X)$ is an integrating factor for (6.18) so

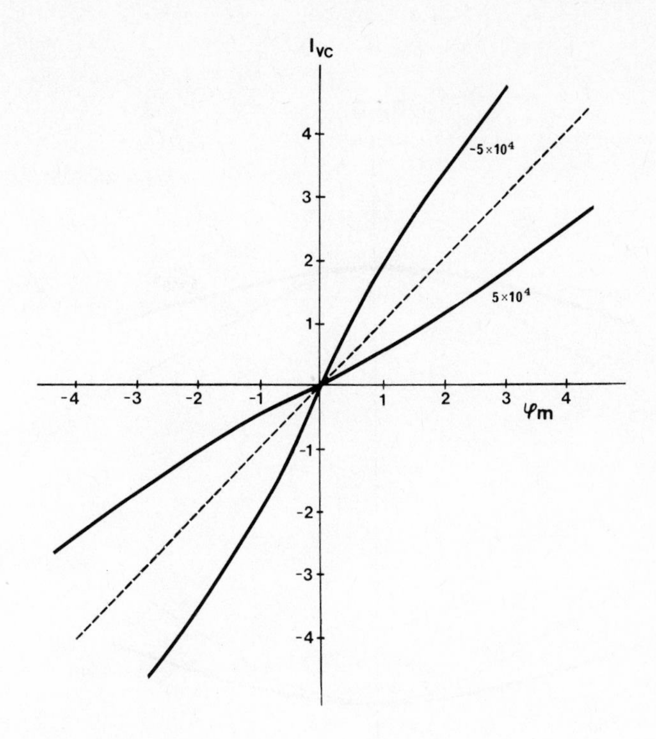

Figure 10.1 Membrane current, I_{VC}, as a function of applied membrane potential, φ_m, as calculated from equation 10.8 with Z = 1, δ =100A (N_i/N_d) = 1, = 3, T = 300°K and volume charge densities of $\pm$ 5 x 10^4 coul/mt^3. The linear Goldman predicted current-membrane potential is shown as a dashed line.

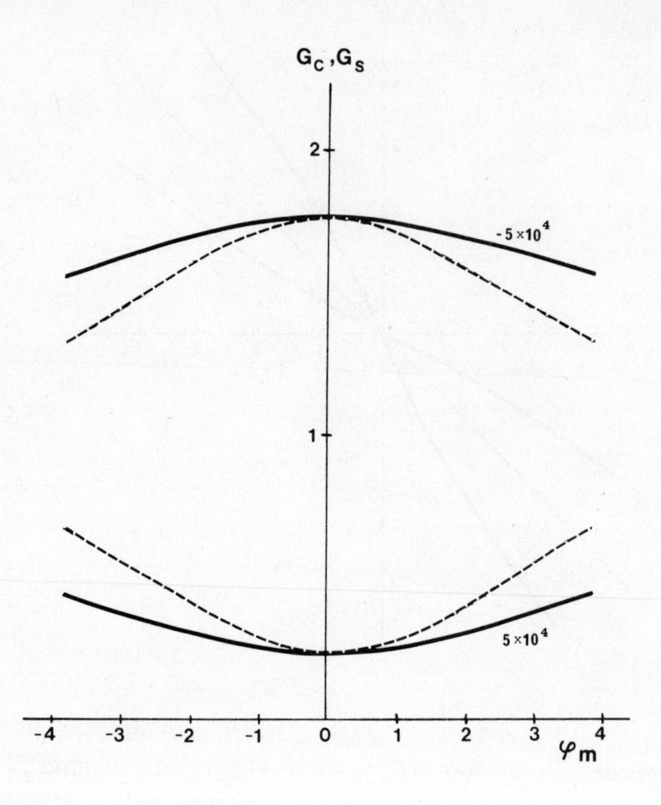

Figure 10.2 Computed chord and slope conductances for the current versus membrane potential relations of Figure 10.1. The chord conductance is shown as a solid line, the slope conductances as dashed lines. The Goldman equation predicts $G_c = G_s = 1$ for all membrane potentials.

$$-I_{VC}\exp\left[Z\varphi(X)\right]dX = Zd\left[N\exp(Z\varphi(X)\right]$$

or with $N=N_i(N_d)$ at $X=0(1)$

$$I_{VC} = Z\frac{N_i\exp c - N_d}{f(a,c)} \qquad (10.8)$$

wherein

$$f(a,c) = \int_0^1 \exp\left[Z\varphi(X)\right] dX \qquad (10.9)$$

is always a real valued function.

It is straightforward, though moderately tedious, to demonstrate that $I_{VC} \sim I_G$ for $|c| >> |a|$. Further, for $Z\bar{\rho} > 0 (< 0)$, $I_{VC} < I_G$ $(> I_G)$ as demonstrated above by use of the mean value theorem.

To illustrate these computed effects of volume fixed charge in the membrane on the current voltage curve I assumed a membrane thickness of 100 A, $\varepsilon = 3$, and $T = 300^{o}K$. Figure 10.1 illustrates the effects of two charge densities on the I_{VC} versus φ_m relation, in the absence of concentration gradients. Figure 10.2 gives the corresponding chord and slope conductance variations, and in both of these figures the expected Goldman result is indicated by the solid line. In 10.3 and 10.4 I show the computed current and conductance curves respectively but with $(N_i/N_d) = 20$, appropriate for K^+ in the squid giant axon. Finally, in Figure 10.5 I show the effect of "titrating" the volume charge on the membrane current-voltage relation. The relationship of these concepts of I versus φ_m modification via volume charge effects to data interpretation remain largely unexplored.

B. Surface Charge Effects. If, instead of a constant volume charge density within the membrane, I assume that there are surface charges at the interior and exterior membrane-electrolyte interfaces, then the potential profile across the membrane can be visualized as in Figure 10.6. A normal Goldman profile will exist within the membrane, given by

$$\varphi(X) = (\varphi_d - \varphi_i)X + \varphi_i \qquad (10.10)$$

while there is an exponential approach from the surface potentials to the potential values in the intra- and extracellular fluid regions

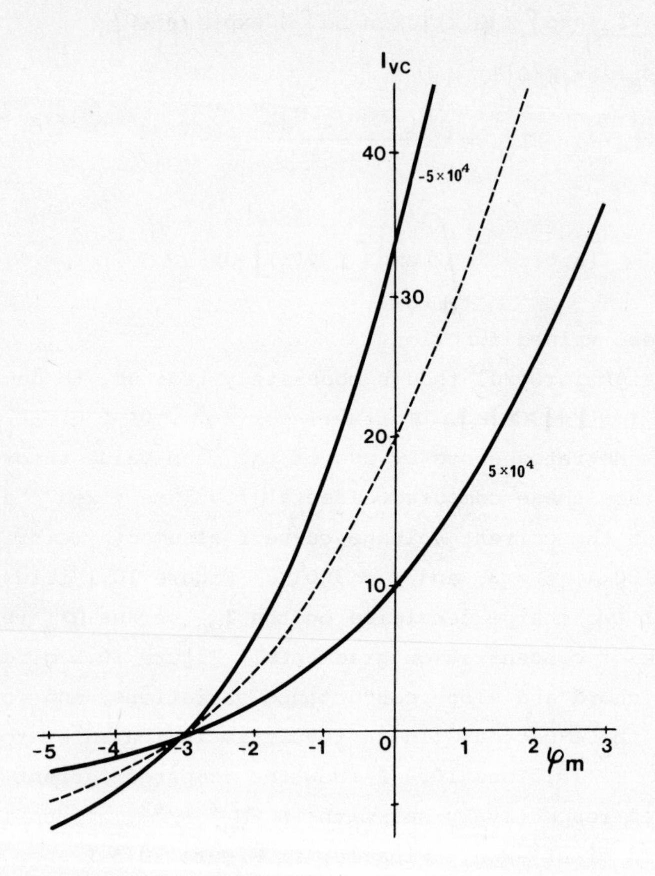

Figure 10.3 Dimensionless current I_{VC} as a function of the dimensionless membrane potential, φ_m, for fixed charge densities of $\pm$ 5 x 10^4 coul/mt^3, (N_i/N_d) = 20, Z = 1 from equation 10.8. The Goldman equation result is shown as a dashed line. All other quantities as in Figure 10.1.

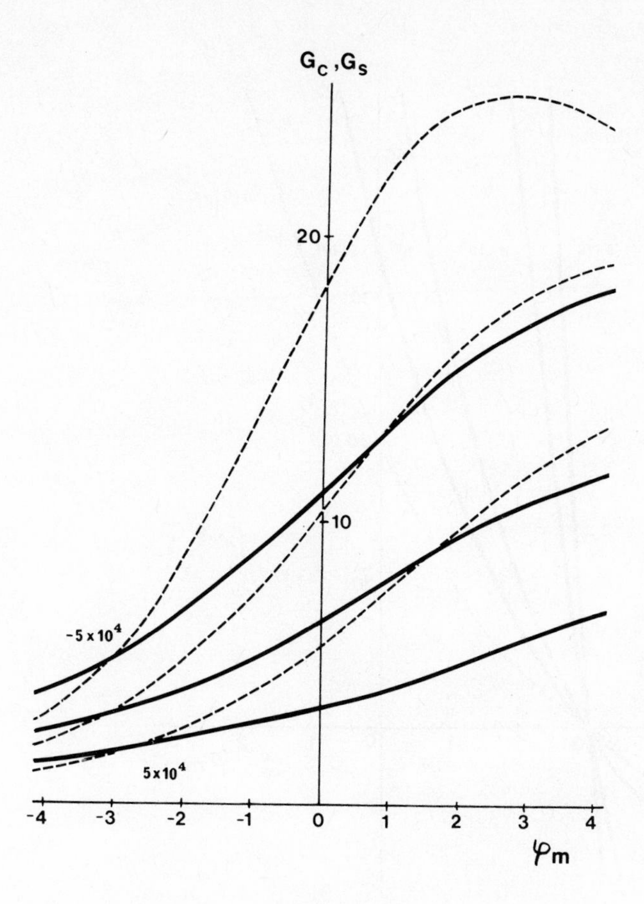

Figure 10.4 Chord and slope conductance versus membrane potential for the current-voltage relations of Figure 10.3. Chord conductances are solid lines, slope conductances are dashed. The Goldman results are presented for comparison.

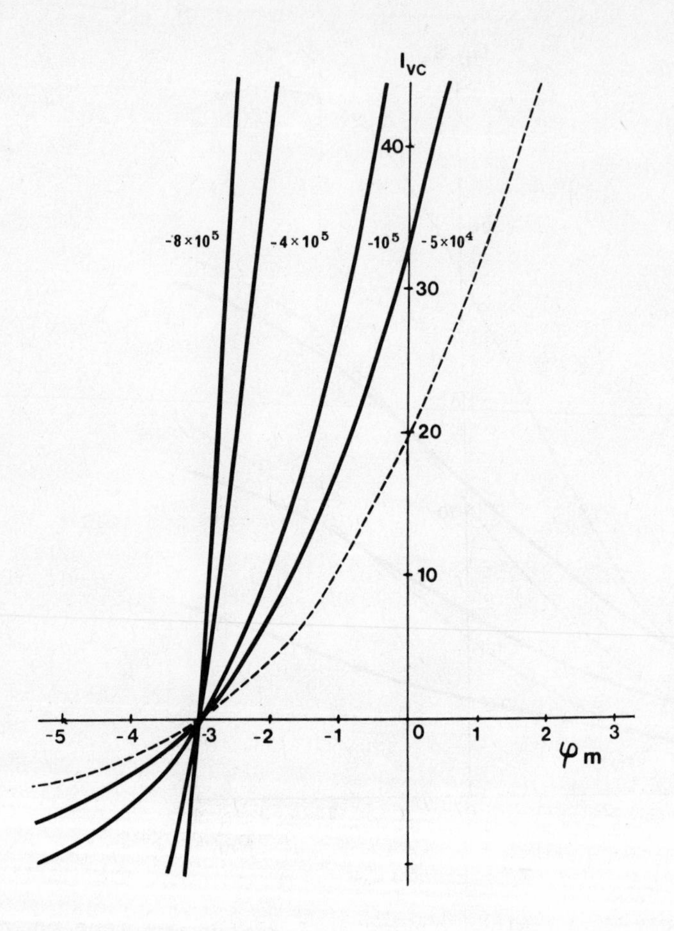

Figure 10.5 I_{VC} versus membrane potential, φ_m, as the volume charge density is altered. All other conditions are as in Figure 10.3.

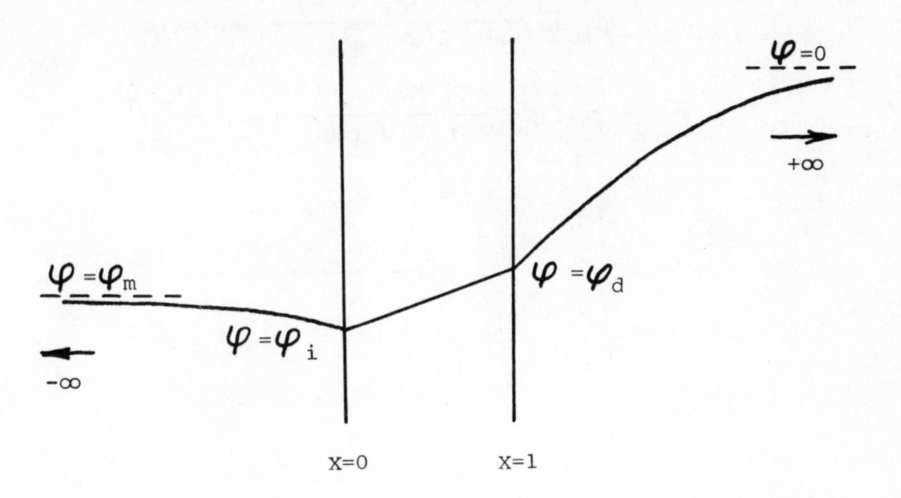

Figure 10.6. Hypothetical potential profile across a membrane that has fixed surface charge at both interior and exterior membrane electrolyte boundaries.

far from the membrane. The existence of the surface potentials will also modify the ionic concentrations at the membrane borders by multiplicative Boltzmann factors, viz.

$$N_{i,0} = N_i \exp\left[Z(\varphi_m - \varphi_i)\right]$$

and (10.11)

$$N_{d,1} = N_d \exp\ Z\varphi_d$$

where N_i and N_d are bulk concentrations as before.

Integrating the Nernst-Planck equation from one interface to the other gives

$$I_{SC} = Z^2(\varphi_i - \varphi_d)\frac{N_{i,0}\exp(Z\varphi_i) - N_{d,1}\exp(Z\varphi_d)}{\exp(Z\varphi_i) - \exp(Z\varphi_d)}$$

or, using (10.11),

$$I_{SC} = Z^2(\varphi_i - \varphi_d)\frac{N_i\exp(Z\varphi_m) - N_d}{\exp(Z\varphi_i) - \exp(Z\varphi_d)} \qquad (10.12)$$

Set $\varphi_m = \varphi_i + \Delta$ and $\bar{\varphi} = \varphi_d + \Delta$ so (10.12) takes the alternate form

$$I_{SC} = Z^2(\varphi_m - \bar{\varphi})\frac{N_i\exp(Z\varphi_m) - N_d}{\exp\ Z(\varphi_m - \bar{\varphi})\ -1}\exp(-Z\varphi_d)$$

$$= I_G\exp(-Z\varphi_d)\frac{\exp(Z\varphi_m) - 1}{\exp\ Z(\varphi_m - \bar{\varphi})\ -1}\cdot\frac{\varphi_m - \bar{\varphi}}{\varphi_m} \tag{10.13}$$

PROBLEMS

10.1. Discuss in detail the analytical properties of I_{VC} as a function of φ_m as given in equation 10.8. Examine the limiting cases of large and small membrane potentials, and determine whether the existence of volume charge in the membrane can ever lead to negative slope conductance regions in the current voltage curve.

10.2. As in the previous problem, but for I_{SC} as given by (10.13).

10.3. Compare and contrast the effects of surface charge and volume charge in the membrane with respect to modification of the membrane current-voltage curves.

10.4. Assume that a membrane has constant volume charge within it and surface charge at both membrane-electrolyte interfaces. Derive an expression for the membrane current as a function of membrane potential, and discuss the properties of the system.

PART III. A MOLECULAR TREATMENT OF TRANSMEMBRANE ION MOVEMENT

In Part II I made exclusive use of the Nernst-Planck formulation of electrodiffusion theory in exploring theoretical explanations of membrane electrical properties. As a heuristic tool this simplistic view is valuable, affording qualitative insights into some of the factors operating that determine steady state and time dependent membrane electrical properties. However, these insights are generally only valid near equilibrium, and in any case completely fail in formulating an understanding of passive sodium movement across the excitable plasma membrane in the steady state.

Thus, there is an immediate reason to search for more detailed models to accurately reflect membrane electrical properties far from equilibrium, and especially to examine possible reasons for the anomolous behaviour of sodium. The material presented in the chapters of this last section are directed toward these goals and although obviously not reaching them it points the way to further research in this area.

The orientation of the models analyzed here is a kinetic theory one and Chapter 11 presents the basic mathematical development of the model. It is long and need not be completely read to follow the other chapters. Reference to the summary section of Chapter 11 will suffice to introduce the reader to following chapters. Chapter 12 shows the formal connection between the two formulations of ED theory. Chapter 13 examines the steady state properties of the kinetic theory model for a membrane system bounded by identical binary electrclytes of equal concentration. The voltage dependence of chord conductance, ionic selectivity and temperature coefficients are treated there. Chapter 14 continues with an examination of the model membrane electrical properties for asymmetric binary electrolyte concentrations (i.e. a concentration gradient) including conductance properties, rectification ratios, and a discussion of the 'independence principle'. Chapter 15 examines the steady state and time dependent properties of the general ED model of Chapter 6.

There is much experimental evidence that ion penetration through excitable cell plasma membranes occurs only at rare specific sites (c.f. Chapter 5). The following assumptions are made about the properties of these sites only, not those of the ion impermeable regions. Some of these assumptions are justifiable from experimental data or order-of-magnitude calculations. Others, however, are not easily justified (or rejected), but are necessary to make initial calculations feasible. These assumptions result in a generalized form of the Nernst-Planck equation, which reduces to the more familiar form at low electric field strengths; hence any objections to these assumptions also apply to the usual Nernst-Planck transport equation. Also, the size of the calculated ionic conductances agree well with experimental data.

It is assumed that the passage of an ion through an ion permeable region is impeded by interactions (collisions) between the ion and membrane molecules, the latter having no directed motion but possessing random thermal motion. These interactions are assumed to be the only ones of importance, implying that ionic densities in the membrane are low. The ion-membrane molecule interactions are assumed to be determined by the following characteristics:

(1) Spherical membrane molecules. The membrane molecules may be represented by spherical particles of finite mass.

(2) Central interactions. The force between an ion and a membrane molecule during a collision is central and conservative. This is related to (1), but carries an added restriction: There are only binary collisions between ions and membrane molecules, i.e., the scatterer centers (collision sites) are widely spaced in comparison to ion diameter.

(3) Small collisional energy loss. During a collision, the

fractional energy loss (ξ) by an ion is much less than one. This is equivalent to the assumption that the mass of a scatterer, compared to the mass of an ion, is large but not infinite (i.e., $\xi \neq 0$). The assumption that $\xi << 1$ greatly simplifies an expansion used in the analysis. Relaxation of this assumption requires the inclusion of more expansion terms, and increases the mathematical complexity of the analysis.

(4) Negligible role of water. This assumption must be made to avoid the complex difficulties of liquid transport theory and may be justified in several ways. Models of ion penetration regions (channels) can be constructed where water would seem to play a negligible role. For example, the wall of a channel having a diameter on the order of a hydrated ion might bind water; but only small clumps of, or single, water molecules would be moving through the channel. Collisions between hydrated ions and single water molecules in the channel would result in little energy loss by the ion because of its much greater mass. Thus, in this situation, water would have little effect on ionic current.. Alternately the effects of water, if present, could be considered to be lumped with the scattering molecules. A final justification is that the kinetic equations derived here are qualitatively identical with those derived by Rice and Gray (1965; eqns. 5.3.69-5.3.71) for transport in liquids.

(5) External forces. All external forces acting on the ion are independent of ionic velocity (e.g., are not due to magnetic fields) and are much smaller than the ion-membrane molecule forces acting during a collision.

CHAPTER 11. MATHEMATICAL FORMULATION OF THE MODEL.

Because the techniques employed in this section are ones generally unfamiliar to membrane biophysicists, although by no means unfamiliar to kinetic theorists, a perhaps disproportionate amount of basic kinetic theory is gathered together from a number of diverse sources as backgouund for the material of later chapters.

For the sake of those unfamiliar with the techniques of non-equilibrium statistical mechanics, mathematical rigor is sacrificed in favor of physical intuition whenever the presentation seemed to benefit. The basic kinetic theory included in this chapter is intended to give only a reading comprehension for what follows, and to those unfamiliar with the field who wish a more detailed presentation texts by Delcroix (1960); Hirschfelder, Curtiss and Bird (1964); and Green (1952) are recommended. Those wishing to skip this chapter may refer to the summary at the end.

A. The Concept of the Distribution Function

In principle, if the position and velocity of every ion or molecule in a mixture at time t = 0 was known, classical mechanics would allow the subsequent motion of every particle to be calculated from the equation of motion:

$$m_i \dot{v}_i = -\nabla_{x_i} H$$

and

$$m_i \dot{x}_i = \nabla_{v_i} H$$

where H is the Hamiltonian of the system defined by

$$H = \sum_i \frac{1}{2} m_i v_i^2 + \Phi(\{x_i\})$$

m_i is the mass of a particle of the i th species, $\underset{\sim}{v}_i$ is the vector velocity of this particle, $\underset{\sim}{x}_i$ is its vector position and $\Phi(\{x_i\})$ the potential energy of the system at a point. However this procedure is not only beyond reasonable solution but also contains information far too fine to be useful. Thus a statistical process is used to describe the properties of a mixture of particles.

In this spirit the following is introduced:

DEFINITION: If $\underset{\sim}{x}$ and $\underset{\sim}{v}_i$ denote vector position and velocity respectively, then the DISTRIBUTION FUNCTION $f_i(\underset{\sim}{x}, \underset{\sim}{v}_i, t)$ for the i th species of a mixture is defined as the number of particles of the i th species which, at a time t, are in a spatial volume element, $dV_{\underset{\sim}{x}}$, about $\underset{\sim}{x}$ and a velocity volume element, $dV_{\underset{\sim}{v}_i}$, about $\underset{\sim}{v}_i$.

As it stands this definition may not be immediately obvious since it requires thought in a seven dimensional phase space (three each for position and velocity, and one for time). Some clarification may result from considering the situation where particles may move in only one dimension. At any time t_o, there are only two dimensions - the position and velocity of a particle on a line. The spatial "volume element" is just the incremental length dx, the velocity "volume element" is dv, and the distribution function f(x,v,t) is the number of particles in the length x_o to x_o+dx whose velocities are between v_o and v_o+dv at time t_o.

B. The Relation of Some Macroscopic Quantities to the Distribution Function

A number of macroscopic quantities such as number density, pressure, mass flux, momentum flux, and energy flux may be formally related to the distribution function.

In terms of the example of the previous section if the total number of particles between x and x+dx is desired, the proper route for this calculation would be to sum up the distribution function for all possible velocities. Mathematically this means that the distribution function should be integrated over the velocity v. In a similar manner for the general case of a distribution function which depends on the vectors $\underset{\sim}{x}$ and $\underset{\sim}{v}_i$, the number density of the i th species can be written as

$$n_i(\underset{\sim}{x},t) = \int f_i(\underset{\sim}{x},\underset{\sim}{v}_i,t)dV_{\underset{\sim}{v}_i}$$

The mass flux of a species is important because it leads to an expression for the current density carried by an ionic species. The mass flux of a particular species is related to the distribution function. Consider first the movement of particles in a three dimensional space - assuming for simplicity that all of the particles are moving in the same direction. Consider a volume element (Figure 11.1) of height $|\underset{\sim}{v}_i|$ dt and cap area dS.

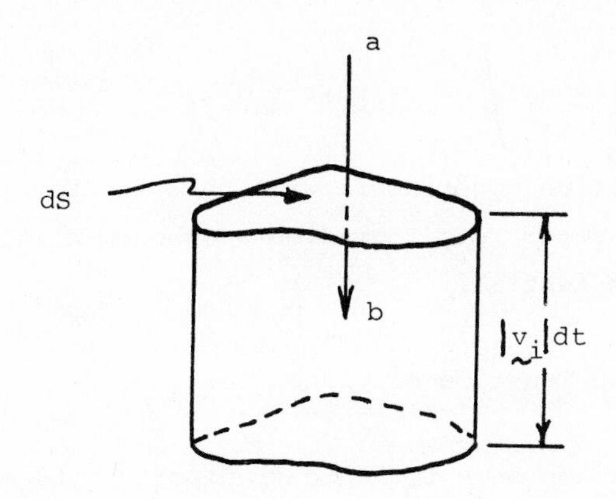

Figure 11.1. Determination of one-dimensional mass flux.

The particles of the i th species are moving in the direction ab, the velocity of one of these particles is $|\underset{\sim}{v}_i|$, and any particle with a velocity between $|\underset{\sim}{v}_i|$ and $|\underset{\sim}{v}_i+d\underset{\sim}{v}_i|$ will cross dS in a time dt if and only if it is in the volume element $dV = |\underset{\sim}{v}_i|dtdS$ at time t=0 The mass of particles of the i th species in dV with velocities between $|\underset{\sim}{v}_i|$ and $|\underset{\sim}{v}_i+d\underset{\sim}{v}_i|$ is $m_i f_i(\underset{\sim}{x},\underset{\sim}{v}_i,t)|\underset{\sim}{v}_i|dtdS$, where m_i is the mass per particle of the i th species. Hence the total mass carried across dS in a time dt by the i th species will be obtained by integrating $m_i f_i|\underset{\sim}{v}_i|dtdS$ over all velocities. Thus the mass flux (mass transported per unit time-unit area) is given by

$$\mathcal{m}_i = \int m_i f_i(x,v_i,t)v_i dv_i$$

In an entirely analogous manner, the mass flux of the i th species of a mixture when motion is not in a single direction is given by

$$\underset{\sim}{\mathcal{M}}_i = \int m_i f_i(\underset{\sim}{x}, \underset{\sim}{v}_i, t)\underset{\sim}{v}_i dV_{\underset{\sim}{vi}}$$

and thus in the general case the mass flux is a vector quantity.

Similarly the number flux, $\underset{\sim}{N}_i = \underset{\sim}{M}_i/m_i$, is given by

$$\underset{\sim}{N}_i = \int f_i(\underset{\sim}{x}, \underset{\sim}{v}_i, t)\underset{\sim}{v}_i, dV_{\underset{\sim}{vi}}$$

If the i th species is charged, and each particle carries a charge $q_i = ez_i$ where z_i is the signed valence of an ion of the i th species, then the current density due to flow of the i th species, is $q\underset{\sim}{N}_i$, or

$$\underset{\sim}{j}_i = q_i \int f_i(\underset{\sim}{x}, \underset{\sim}{v}_i, t)\underset{\sim}{v}_i dV_{\underset{\sim}{vi}}$$

In general the distribution function f_i will be a function of various driving forces in the system, e.g., gradients of concentration, temperature, or electric potential.

C. The Derivation of an Equation of Change for the Distribution Function

If this statistical approach to ionic transport is to be of any value, the distribution function must be expressed as a function of velocity so that the integrations of section B may be performed. In this section an equation of change for f_i will be derived, whose solution will be considered later.

Consider a point in space located by a vector $\underset{\sim}{x}$ which has a small volume element $dV_{\underset{\sim}{x}}$ around it, and also a point in velocity space characterized by a vector $\underset{\sim}{v}_i$ and surrounded by a volume element $dV_{\underset{\sim}{vi}}$. The number of particles of the i th species which are in $dV_{\underset{\sim}{x}}$ and whose velocities are in $dV_{\underset{\sim}{vi}}$ at a time t will then be $f_i(\underset{\sim}{x}, \underset{\sim}{v}_i, t)dV_{\underset{\sim}{x}}dV_{\underset{\sim}{vi}}$. Assume that there are no collisions between particles and that an external force $\underset{\sim}{F}_i$ is acting on the i th species. After a short time dt the particles in the volume element $dV_{\underset{\sim}{x}}$ will have moved a small distance $\underset{\sim}{v}_i dt$ to a new volume $dV_{\underset{\sim}{x}}'$ located at $\underset{\sim}{x}+\underset{\sim}{v}_i dt$, and the particles whose velocities were in $dV_{\underset{\sim}{vi}}$ will have moved to a new volume $dV_{\underset{\sim}{v_i}}'$ in velocity space located at $\underset{\sim}{v}_i+(\underset{\sim}{F}_i/m_i)dt$. Thus with no collisions the relation expressing conservation of particles is

$$f_i(\underset{\sim}{x},\underset{\sim}{v}_i,t)dV_{\underset{\sim}{x}}dV_{\underset{\sim}{v_i}} = f_i(\underset{\sim}{x}+\underset{\sim}{v}_i dt,\underset{\sim}{v}_i+\frac{\underset{\sim}{F}_i}{m_i}dt,t+dt)\ dV_{\underset{\sim}{x}}'dV_{\underset{\sim}{v_i}}' \qquad (11.1)$$

If collisions occur some of the particles of the i th species found in $dV_{\underset{\sim}{x}}'$ and $dV_{\underset{\sim}{v_i}}'$ at time t+dt may not have been in $dV_{\underset{\sim}{x}}$ and $dV_{\underset{\sim}{v_i}}$ at time t. Similarly, there may be particles of the i th species which were in $dV_{\underset{\sim}{x}}$ with their velocities in $dV_{\underset{\sim}{v_i}}$ at time t that never made it to $dV_{\underset{\sim}{x}}'$ and $dV_{\underset{\sim}{v_i}}'$ at time t+dt because of collisions. Collisions are accounted for in the following way:

DEFINITIONS: Let $\Delta^l_{ij}(\Delta^g_{ij})$ be the number density of particles of the i th species lost from (gained by) a small spatial volume $dV_{\underset{\sim}{x}}(dV_{\underset{\sim}{x}}')$ and a small volume in velocity space $dV_{\underset{\sim}{v_i}}(dV_{\underset{\sim}{v_i}}')$ in a time dt due to collisions with particles of the j th species.

Hence the number of particles of the i th species lost from (gained by) the volume elements $dV_{\underset{\sim}{x}}$ and $dV_{\underset{\sim}{v_i}}(dV_{\underset{\sim}{x}}'$ and $dV_{\underset{\sim}{v_i}}')$ due to collisions with particles of the j th species in a time dt is given by $\Delta^l_{ij}dV_{\underset{\sim}{x}}dV_{\underset{\sim}{v_i}}dt(\Delta^g_{ij}dV_{\underset{\sim}{x}}'dV_{\underset{\sim}{v_i}}'dt)$. Therefore if collisions occur the previous balance equation 11.1 should be replaced by

$$f_i(\underset{\sim}{x}+\underset{\sim}{v}_i dt,\ \frac{1}{m_i}\underset{\sim}{F}_i dt+\underset{\sim}{v}_i,\ t+dt)dV_{\underset{\sim}{x}}'dV_{\underset{\sim}{v_i}}' = f_i(\underset{\sim}{x},\underset{\sim}{v}_i,t)dV_{\underset{\sim}{x}}dV_{\underset{\sim}{v_i}}$$

$$+\sum_j\left[\Delta^g_{ij}dV_{\underset{\sim}{x}}'dV_{\underset{\sim}{v_i}}' - \Delta^l_{ij}dV_{\underset{\sim}{x}}dV_{\underset{\sim}{v_i}}\right]dt \qquad (11.2)$$

The volume elements at times t and t+dt in equation 11.2 are related through their Jacobian by

$$dV_{\underset{\sim}{x}} = J_{\underset{\sim}{x}}(\underset{\sim}{x},\ \underset{\sim}{x}+\underset{\sim}{v}_i dt)dV_{\underset{\sim}{x}}'$$

where the Jacobian is defined by

$$J(p,q) = \frac{\partial q}{\partial p}$$

Thus $J_{\underset{\sim}{x}} = 1$ so $dV_{\underset{\sim}{x}} = dV_{\underset{\sim}{x}}'$, and it may also be shown that $dV_{\underset{\sim}{x}}dV_{\underset{\sim}{v_i}} = dV_{\underset{\sim}{x}}'dV_{\underset{\sim}{v_i}}'$ so the balance equation 11.2 becomes

$$f_i(\underset{\sim}{x}+\underset{\sim}{v}_i dt,\ \underset{\sim}{v}_i+\frac{1}{m_i}\underset{\sim}{F}_i dt,t+dt) =$$

$$f_i(\underset{\sim}{x},\underset{\sim}{v}_i,t) + \sum_j\left[\Delta^g_{ij} - \Delta^l_{ij}\right]dt \qquad (11.3)$$

The Taylor series expansion of a function $G(\underset{\sim}{a}+\underset{\sim}{g},\underset{\sim}{b}+\underset{\sim}{h},c+d)$ about $(\underset{\sim}{a},\underset{\sim}{b},c)$ is given by

$$G(\underset{\sim}{a},\underset{\sim}{b},c) + \sum_{n=1}^{\infty} \frac{1}{n!}\left[\left(\underset{\sim}{g}\cdot\nabla_{\underset{\sim}{x}}+\underset{\sim}{h}\cdot\nabla_{\underset{\sim}{y}}+d\frac{\partial}{\partial t}\right)^{n} G(\underset{\sim}{x},\underset{\sim}{y},t)\right] ,$$

where $\underset{\sim}{x}$ and $\underset{\sim}{y}$ are dummy variables, so the left hand side of (11.3) may be written as

$$f_i(\underset{\sim}{x}+\underset{\sim}{v}_i dt,\underset{\sim}{v}_i+\frac{1}{m_i}\underset{\sim}{F}_i dt,t+dt) = f_i(\underset{\sim}{x},\underset{\sim}{v}_i,t) +$$

$$\left[\underset{\sim}{v}_i\cdot\nabla_{\underset{\sim}{x}} f_i+\frac{1}{m_i}\underset{\sim}{F}_i\cdot\nabla_{\underset{\sim}{v}_i} f_i + \frac{\partial f_i}{\partial t}\right] dt + O(dt^2) \qquad (11.4)$$

Substituting (11.4) into (11.3) and neglecting terms of $O(dt^2)$ gives the balance equation in a slightly different form:

$$\frac{\partial f_i}{\partial t} + \underset{\sim}{v}_i\cdot\nabla_{\underset{\sim}{x}} f_i + \frac{1}{m_i}\underset{\sim}{F}_i\cdot\nabla_{\underset{\sim}{v}_i} f_i = \sum_j \left[\Delta^{g}_{ij} - \Delta^{l}_{ij}\right] \qquad (11.5)$$

This equation of change for the distribution function f is also known as the Boltzmann equation. The collision terms on the right hand side of the equation, usually denoted by $\delta f_i/\delta t$, may be written more specifically in terms of distribution functions and parameters characterizing the collision process. However, it is first necessary to discuss collision processes.

D. The Dynamics of a Binary Collision

This section gives a discussion of a collision process between two particles based on classical physics. Here a collision is said to occur in an interval t_o to t_1 if the potential energy of interaction between two particles varies at all times t in the interval from the value at infinite separation of the two particles.

Interparticle forces are assumed to be central and conservative and are therefore derivable from a symmetric potential energy function $\varphi(r)$. It is assumed that these are the only forces acting during a collision.

If particles of the i th and j th molecular species are located in space by the vectors $\underset{\sim}{x}_i$ and $\underset{\sim}{x}_j$ respectively, then the forces on the i th particle due to the j th is

$$\underset{\sim}{\mathcal{F}}_i = -\nabla_{\underset{\sim}{x}_i} \varphi(r_{ji})$$

while the forces on the j th particle due to the i th is

$$\underset{\sim}{\mathcal{F}}_i = -\nabla_{\underset{\sim}{x}_j} \varphi(r_{ji})$$

where $r_{ji} = |\underset{\sim}{r}_{ji}| = |\underset{\sim}{x}_j - \underset{\sim}{x}_i|$. Alternately these two equations may be written

$$\underset{\sim}{\mathcal{F}}_i = \nabla_{\underset{\sim}{r}_{ji}} \varphi(r_{ji}) = -\mathcal{F}_j$$

The relative velocity, $\underset{\sim}{g}_{ji}$, of a particle of the j th species with respect to a particle of the i th is given by

$$\underset{\sim}{g}_{ji} = \underset{\sim}{v}_j - \underset{\sim}{v}_i = \dot{\underset{\sim}{x}}_j - \dot{\underset{\sim}{x}}_i,$$

so the relative acceleration has the form

$$\dot{\underset{\sim}{g}}_{ji} = \frac{1}{m_j}\underset{\sim}{\mathcal{F}}_j - \frac{1}{m_i}\underset{\sim}{\mathcal{F}}_i = -\frac{1}{\mu_{ij}}\nabla_{\underset{\sim}{r}_{ji}} \varphi(r_{ji}) \tag{11.6}$$

where

$$\mu_{ij} = \frac{m_i m_j}{m_i + m_j}$$

is defined as the reduced mass of the two particles. Equation 11.6 for the relative acceleration of the two particles is identical to the acceleration of a particle of mass μ_{ij} in a central potential field $\varphi(r_{ji})$. Thus instead of treating the movement of the two particles, the analogous problem of a single particle in a central force field may be considered.

The angular momentum of this equivalent particle is given by

$$\underset{\sim}{L}_{ji} = \underset{\sim}{r}_{ji} \times (\mu_{ij}\underset{\sim}{g}_{ji})$$

and simple vector manipulations show that

$$\frac{\partial L_{ji}}{\partial t} = 0 \tag{11.7}$$

so angular momentum is conserved during the collision. From equation 11.7 the relation

$$\underset{\sim}{r}_{ji} \times \underset{\sim}{g}_{ji} = \underset{\sim}{A}$$

where the vector $\underset{\sim}{A}$ is independent of time, must be true and therefore during a collision the vectors $\underset{\sim}{r}_{ji}$ and $\underset{\sim}{g}_{ji}$ lie in a fixed plane. Thus during a collision the motion of the equivalent particle may be visualized as in Figure 11.2.

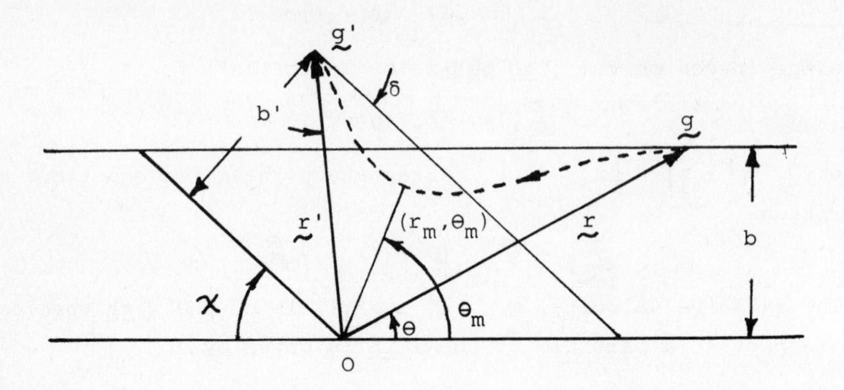

Figure 11.2. Motion of an equivalent particle in a central force field.

In this figure all unprimed quantities refer to pre-collisional values while primed quantities are post-collisional. The quantity b_{ij} is known as the impact parameter and is the closest the reduced particle would come to the point 0 in the absence of the potential $\varphi(r_{ij})$. The point (r_m, Θ_m) is the point of closest approach in the presence of $\varphi(r_{ij})$, and the angle χ_{ij} is called the scattering angle.

The energy of the equivalent particle is

$$E_{ij} = \frac{1}{2}\mu_{ij}g_{ij}^2 + \varphi(r_{ji})$$

so

$$\frac{\partial E_{ij}}{\partial t} = 0$$

and energy is also conserved during the collision. Using this fact, and requiring that $\varphi(r_{ij}) \rightarrow 0$ as $r_{ij} \rightarrow \infty$, it is clear that in the pre- and post-collisional states the relation

$$\frac{1}{2}\mu_{ij}g_{ij}^2 = \frac{1}{2}\mu_{ij}(g_{ij}')^2$$

or

$$g_{ij} = g_{ij}' \tag{11.8}$$

must hold.

From the conservation of angular momentum,

$$|\underset{\sim}{r}_{ji} \times \underset{\sim}{g}_{ji}| = |\underset{\sim}{r}_{ji}' \times \underset{\sim}{g}_{ji}|$$

and with (11.8) this reduces to

$$r_{ji} \sin \Theta = r'_{ji} \sin \delta$$

From the geometry of the motion depicted in Figure 11.2 it is clear that this is equivalent to

$$b_{ji} = b'_{ji}$$

Before concluding this discussion the following definition is introduced:

DEFINITION: The differential scattering cross section, I_{ij}, for the j th species is defined as the number of particles of the i th species scattered or deflected into the solid angle $d\Omega_{ij} = 2\pi \sin\chi_{ij} d\chi_{ij}$ per unit time, divided by the pre-collisional number flux, η_i, of the i th species.

The number of particles of the i th species deflected through an angle between χ_{ij} and $\chi_{ij} + d\chi_{ij}$ is identical with the number of particles which, in the pre-collisional state, had impact parameters between b_{ij} and $b_{ij} + db_{ij}$. However, this number is just $2\pi \eta_i b_{ij} db_{ij}$ so from the definition of I_{ij},

$$I_{ij} d\Omega_{ij} = 2\pi b_{ij} db_{ij} \tag{11.9}$$

E. Evaluation of the Boltzmann Equation Collision Term

The concepts of the previous section can be used to evaluate the collision terms in equation 11.5. Suppose a particle of the i th species is located as pictured in Figure 11.3 and a particle of the j th species is approaching with a relative velocity g_{ji}. The num-

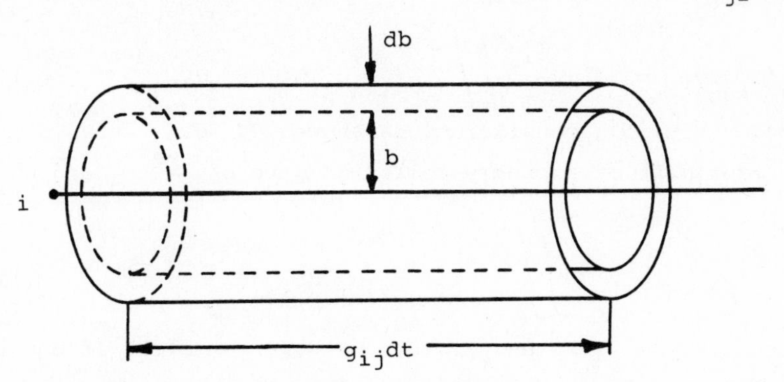

Figure 11.3. Collisions of particles of the j th species with one of the i th.

ber of particles of the j th species in the cylindrical shell depicted in Figure 11.3 will be $2\pi f_j g_{ji} b_{ij} db_{ij} dt$, so the total number of collisions between particles of the j th species and the fixed i th particle will be

$$2\pi \int f_i(\underset{\sim}{x},\underset{\sim}{v}_j,t) g_{ji} b_{ij} db_{ij} dV_{\underset{\sim}{v}j}$$

The probable number of particles of the i th species to be found in a volume $dV_{\underset{\sim}{x}}$ about $\underset{\sim}{x}$ and $dV_{\underset{\sim}{v}i}$ about $\underset{\sim}{v}_i$ is $f_i(\underset{\sim}{x}_i,\underset{\sim}{v}_i,t)dV_{\underset{\sim}{x}}dV_{\underset{\sim}{v}i}$. Hence the number of particles of the i th species removed from $dV_{\underset{\sim}{x}}dV_{\underset{\sim}{v}i}$ by collisions with particles of the j th species in a time dt (i.e., $\Delta^{l}_{ij}dV_{\underset{\sim}{x}}dV_{\underset{\sim}{v}i}dt$) will be given by

$$\Delta^{l}_{ij}dV_{\underset{\sim}{x}}dV_{\underset{\sim}{v}i} = 2\pi dV_{\underset{\sim}{x}}dV_{\underset{\sim}{v}i}dt \iint f_i f_j g_{ji} b_{ij} db_{ij} dV_{\underset{\sim}{v}j},$$

or

$$\Delta^{l}_{ij} = 2\pi \iint f_i f_j g_{ji} b_{ij} db_{ij} dV_{\underset{\sim}{v}j} \tag{11.10}$$

In the foregoing determination of Δ^{l}_{ij} the j th particle had a collisional relative velocity $\underset{\sim}{g}_{ji}$ and a post-collisional relative velocity $\underset{\sim}{g}'_{ji}$. Because of the central, conservative nature of the inter-particle force the collision may be "run backward" so the particle of the j th species has pre- and post-collisional relative velocities $\underset{\sim}{g}'_{ji}$ and $\underset{\sim}{g}_{ji}$ respectively, and the dynamics of ths situation will be unchanged. Using much the same method as was used in evaluating Δ^{l}_{ij}, the expression

$$\Delta^{g}_{ij}dV_{\underset{\sim}{v}i} = 2\pi dV'_{\underset{\sim}{v}i} \iint f'_i f'_j g'_{ji} b'_{ij} db'_{ij} dV_{\underset{\sim}{v}j} \tag{11.11}$$

is obtained. Previous considerations showed $dV_{\underset{\sim}{v}i}dV_{\underset{\sim}{v}j} = dV'_{\underset{\sim}{v}i}dV'_{\underset{\sim}{v}j}$ while the treatment of a binary collision gave $b_{ij} = b'_{ij}$ and $g_{ji} = g'_{ji}$. Thus (11.11) becomes

$$\Delta^{g}_{ij} = 2\pi \iint f'_i f'_j g_{ij} b_{ij} db_{ij} dV_{\underset{\sim}{v}j} \tag{11.12}$$

Utilizing equations 11.10 and 11.12, and equation 11.9, in equation 11.5 the collision term of the Boltzmann equation becomes

$$\frac{\delta f_i}{\delta t} = \sum_j (\Delta^{g}_{ij} - \Delta^{l}_{ij})$$

$$= \iint \sum_j (f_i' f_j' - f_i f_j) g_{ij} I_{ij} dV_{\underset{\sim}{v_j}} d\Omega \qquad (11.13)$$

In dealing with a system consisting of, say, N different components it is necessary to solve simultaneously N integro-differential equations like (11.13).

F. Simplification of the Boltzmann Equation

1. General remarks. The Boltzmann equation is more tractable if the scatterer possesses no pre-collisional energy (i.e., is truly fixed) and the ion loses no energy during the collision. These assumptions are clearly unphysical, and correction is made later.

If the subscripts i and j denote ion and scatterer respectively these assumptions imply that

$$v_j = v_j' = 0$$

and

$$g_{ij} = v_i = v_i'$$

Thus during an encounter the ionic velocity changes in direction but not in magnitude. The collision term, $\delta f_i / \delta t$, of the Boltzmann equation 11.13 is simplified by these assumptions to

$$\frac{\delta f_i}{\delta t} = \sum_j n_j \int (f_i' - f_i) v_i I_{ij} d\Omega$$

where n_j is the number density of scatterers.

One method of solving the Boltzmann equation is to expand the distribution function, f_i, in a series whose first few coefficients are sufficient to calculate the macroscopic variables of interest; e.g., current. The method used here is to expand the distribution function in a series of orthogonal functions whose orthogonality may be usefully employed. Substitution into the Boltzmann equation, multiplication of the resulting equation by each of the orthogonal functions and integration over an appropriate variable leads to an infinite set of coupled equations. The equation of n th order is coupled to the ones of order (n - 1) and (n +1). This set is terminated arbitrarily and the resultant finite set hopefully solved.

The form of the expansion is suggested by expressing the ion velocity in spherical coordinates (v_i, θ, ϕ), where θ and ϕ are now the variables during a collision, and expressing the distribution function as

$$f_i(\underset{\sim}{x}, \underset{\sim}{v_i}, t) = f_i(x; v_i, \Theta, \phi; t)$$

Expanding in a series of spherical harmonics was first employed by Lorentz (1952) and is used here. The mathematics in the next few paragraphs is tedious, and not especially illuminating in terms of the physics of the membrane model and the material up to equation 11.23 can be skipped without loss of continuity.

2. Expansion of the Distribution Function and its Substitution in Boltzmann's Equation. The distribution function, f(drop the subscript denoting species temporarily), is expanded in a series

$$f = \sum_{p,q,r} f_{pqr} R_{p,q,r} \tag{11.14}$$

where $f_{p,q,r}$ are functions of position, time, and velocity magnitude only, and

$$R_{p,q,r} = P_p^q(\cos\Theta)\left[\delta_{0,r}\cos q\phi + \delta_{1,r}\sin q\phi\right]$$

are the general spherical harmonics. $P_p^q(\cos\Theta)$ is the Legendre function. The first few spherical harmonics are

$(p,q,r) = (0,0,0)$	$(p,q,r) = (1,0,0)$
$R_{0,0,0} = 1$	$R_{1,0,0} = \cos\Theta = \cos\Theta_z$
	$R_{1,1,0} = \sin\Theta\cos\phi = \cos\Theta_x$
	$R_{1,1,1} = \sin\Theta\sin\phi = \cos\Theta_y$

where the $\cos\Theta_x$, etc., are direction cosines. Hence the first few terms in the expansion (11.14) are

$$\begin{aligned} f &= f_{000} + f_{100}\cos\Theta + f_{110}\sin\Theta\cos\phi + f_{111}\sin\Theta\sin\phi \\ &= f_{000} + f_{100}\cos\Theta_z + f_{110}\cos\Theta_x + f_{111}\cos\Theta_y \end{aligned} \tag{11.15}$$

The form of (11.15) immediately suggests the alternative expression

$$f = f_0 + \underset{\sim}{f_1} \bullet \frac{\underset{\sim}{v}}{|\underset{\sim}{v}|} + \ldots \tag{11.16}$$

where

$$f_0 = f_{000}$$

and

$$\underset{\sim}{f_1} = \underset{\sim}{i_x} f_{110} + \underset{\sim}{i_y} f_{111} + \underset{\sim}{i_z} f_{100} \tag{11.17}$$

Indeed, Johnston (1960) has shown that the expansion (11.14) is formally identical with a Cartesian tensor scalar product expansion

$$f = \sum_p \frac{\{f_p\} \vdots_p \{\underset{\sim}{v}^p\}}{v^p}$$

where $\{f_p\}$ is a symmetric p th order Cartesian tensor and

$$\{\underset{\sim}{v}^p\} = \underset{\sim}{v}\,\underset{\sim}{v}\,\underset{\sim}{v} \ldots \underset{\sim}{v} \quad \text{(p times)}$$

Equation 11.16 is a more convenient form of the expansion 11.14 than (11.15), and is the one that is substituted in the Boltzmann Equation. Here, it is assumed that terms in the expansion 11.16 for $p \geq 2$ are negligible and thus the two terms in 11.16 are the only ones used. Morrone (1968) has demonstrated that the expansion (11.16) of f is equivalent to an expansion in ξ. In an expansion in ξ the coefficient of f_j is $(\xi)^{j/2}$ and in keeping with assumption 3 in the introduction to Part III that $\xi \ll 1$ neglect of terms for $j \geq 2$ is justified.

Now substitute equation 11.16 into Boltzmann's equation

$$\frac{\partial f}{\partial t} + \underset{\sim}{v} \cdot \nabla_{\underset{\sim}{x}} f + \frac{1}{m}\underset{\sim}{F} \cdot \nabla_{\underset{\sim}{v}} f = \sum_j n_j \int (f_i' - f) v I_j d\Omega \qquad (11.18)$$

and consider each term separately.

a. $\partial f/\partial t$:

$$\frac{\partial f}{\partial t} = \frac{\partial f_0}{\partial t} + \frac{\partial \underset{\sim}{f}_1}{\partial t} \cdot \frac{\underset{\sim}{v}}{v} + \ldots$$

b. $\underset{\sim}{v} \cdot \nabla_{\underset{\sim}{x}} f$:

$$\underset{\sim}{v} \cdot \nabla_{\underset{\sim}{x}} f = \underset{\sim}{v} \cdot \nabla_{\underset{\sim}{x}} f_0 + \underset{\sim}{v} \cdot \nabla_{\underset{\sim}{x}} \left(\frac{\underset{\sim}{v} \cdot \underset{\sim}{f}_1}{v}\right) + \ldots$$

$$= \underset{\sim}{v} \cdot \nabla_{\underset{\sim}{x}} f_0 + \frac{\underset{\sim}{v}\underset{\sim}{v}}{v} : \nabla_{\underset{\sim}{x}} \underset{\sim}{f}_1 + \ldots.$$

where $\underset{\sim}{v}\underset{\sim}{v}$ is a tensor, as is $\nabla_{\underset{\sim}{x}} \underset{\sim}{f}_1$, and ":" denotes a tensor scalar product;

c. $(\underset{\sim}{F}/m) : \nabla_{\underset{\sim}{v}} f$: This term is more difficult to evaluate and use must be made of the identity $\nabla_{\underset{\sim}{v}} \underset{\sim}{v} = \underset{\approx}{I}$, where $\underset{\approx}{I}$ is the unity tensor. Thus

$$\frac{1}{m}\underset{\sim}{F} \cdot \nabla_{\underset{\sim}{v}} f = \frac{1}{m} \frac{\underset{\sim}{F} \cdot \underset{\sim}{v}}{v} \frac{\partial f_0}{\partial v} + \frac{1}{m} \frac{\underset{\sim}{F} \cdot \underset{\sim}{v}}{v} \frac{\partial}{\partial v}\left(\frac{\underset{\sim}{f}_1}{v}\right) \cdot \underset{\sim}{v} + \frac{\underset{\sim}{F} \cdot \underset{\sim}{f}_1}{mv} + \ldots$$

$$= \frac{1}{m} \frac{\underset{\sim}{F} \cdot \underset{\sim}{f}_1}{v} + \frac{1}{m} \frac{\underset{\sim}{F} \cdot \underset{\sim}{v}}{v} \frac{\partial f_0}{\partial v} + \frac{1}{m}\underset{\sim}{F} \frac{\partial}{\partial v}\left(\frac{\underset{\sim}{f}_1}{v}\right) : \frac{\underset{\sim}{v}\underset{\sim}{v}}{v^2} + \ldots.$$

d. $\delta f/\delta t$: from (11.18)

$$\frac{\delta f}{\delta t} = \sum_j n_j \int \left[f(\underset{\sim}{x},\underset{\sim}{v},\theta',\phi',t) - f(x,v,\theta,\phi,t) \right] vI_j d\Omega \quad (11.19)$$

$$= \sum_j n_j v \sum_{p,q,r} \int f_{pqr} \left[R_{p,q,r}(\theta',\phi') - R_{p,q,r}(\theta,\phi) \right] I_j d\Omega$$

The evaluation of the integral in (11.19) is not dependent on the scattering particle so drop the j subscript temporarily. Expand the differential scattering cross section as a series of Legendre functions,

$$I(v,\chi) = \sum_s I_s(v) P_s(\cos\chi)$$

and substitute it in the integral of (11.19) to give, after some use of the orthogonality relations between spherical harmonics and Legendre functions,

$$\int \left[R_{p,q,r}(\theta',\phi') - R_{p,q,r}(\theta,\phi) \right] I(v,\chi) d\Omega$$
$$= - R_{p,q,r} \int (1 - P_p \cos\chi) I(v,\chi) d\Omega$$

Thus, the collision term becomes

$$\frac{\delta f}{\delta t} = - \sum_j n_j v \sum_{p,q,r} f_{pqr} R_{p,q,r} \int (1-P_p \cos\chi) I_j(v,\chi) d\Omega \quad (11.20)$$

It is customary to define a p <u>th</u> order <u>collision</u> <u>frequency</u>, ν_p, as

$$\nu_p(v) = \sum_j n_j v \int (1-P_p \cos\chi) I_j(v,\chi) d\Omega \quad (11.21)$$

so the collision term (11.20) becomes

$$\frac{\delta f}{\delta t} = - \sum_p \nu_p(v) \sum_{q,r} f_{pqr} R_{p,q,r}$$

or using the ideas expressed in (11.16)

$$\frac{\delta f}{\delta t} = -(\nu_0 f_0 + \nu_1 \underset{\sim}{f_1} \cdot \frac{\underset{\sim}{v}}{v} + \nu_2 \underset{\approx}{f_2} : \frac{\underset{\sim}{v}\underset{\sim}{v}}{v^2} + \ldots) \quad (11.22)$$

Note, however, that from the definition of $\nu_p(v)$, $\nu_0 = 0$ with the simplifying assumptions made above.

3. Corrections for Collisional Energy Loss and Non-Zero

Scatterer Energy. This section describes how corrections can be made for energy loss by an ion during an collision and the thermal motion of the scattering centers. Energy loss will be treated first using the method followed by Morse, Allis and Lamar (1935).

Equation 11.13 may be written in the form

$$\frac{\delta f_i}{\delta t} dV_{\underset{\sim}{v_i}} = \sum_j \iint \Big[f_i' f_j' g_{ij}' I_{ij}' dV_{\underset{\sim}{v_i}}' dV_{\underset{\sim}{v_j}}' - f_i f_j g_{ij} I_{ij} dV_{\underset{\sim}{v_i}} dV_{\underset{\sim}{v_j}} \Big] d\Omega \tag{11.23}$$

if use is not made of the relationship $dV_{\underset{\sim}{v_i}} dV_{\underset{\sim}{v_j}} = dV_{\underset{\sim}{v_i}}' dV_{\underset{\sim}{v_j}}'$. Assuming as before no motion of the scattering particle, 11.23 becomes

$$\frac{\delta f_i}{\delta t} dV_{\underset{\sim}{v_i}} = \sum_j n_j \int \Big[f_i' v_i' I_{ij}' dV_{v_i}' - f_i v_i I_{ij} dV_{\underset{\sim}{v_i}} \Big] d\Omega \tag{11.24}$$

This time, however, the assumption that the particle of the i th species loses no energy during the collision is not made so $v_i \neq v_i'$ and $dV_{\underset{\sim}{v_i}} \neq dV_{\underset{\sim}{v_i}}'$, and v_i must therefore be related to v_i'.

Consider the scattering particle located at $\underset{\sim}{x_j}$ in position space and an ion located at $\underset{\sim}{x_i}$. Then the position of the center of mass, $\underset{\sim}{X_{cm}}$, is given by

$$\underset{\sim}{X_{cm}} = \frac{m_i \underset{\sim}{x_i} + m_j \underset{\sim}{x_j}}{m_i + m_j}$$

Also the velocity of the center of mass, $\underset{\sim}{V_{cm}} = \dot{\underset{\sim}{X}}_{cm}$, is

$$\underset{\sim}{V_{cm}} = \frac{m_i \underset{\sim}{v_i} + m_j \underset{\sim}{v_j}}{m_i + m_j}$$

In a center of mass coordinate system the scatterer and the ion would have velocities relative to the center of mass $\underset{\sim}{v_{j,cm}}$ and $\underset{\sim}{v_{i,cm}}$ respectively. These are related to the original rest velocities by

$$\underset{\sim}{v_j} = \underset{\sim}{V_{cm}} + \underset{\sim}{v_{j,cm}}$$

and

$$\underset{\sim}{v_i} = \underset{\sim}{V_{cm}} + \underset{\sim}{v_{i,cm}}$$

respectively. Thus the change in energy of an ion during a collision is

$$\frac{1}{2}m_i v_i^{\,2} - \frac{1}{2}m_i v_i'^{\,2}$$

$$= \frac{1}{2}m_i \left[(\underset{\sim}{V}_{cm} + \underset{\sim}{v}_{i,cm}) \cdot (\underset{\sim}{V}_{cm} + \underset{\sim}{v}_{i,cm}) - (\underset{\sim}{V}_{cm} + \underset{\sim}{v}'_{i,cm}) \cdot (\underset{\sim}{V}_{cm} + \underset{\sim}{v}'_{i,cm}) \right]$$

$$= m_i \underset{\sim}{V}_{cm} \cdot (\underset{\sim}{v}_i - \underset{\sim}{v}'_i)$$

For zero scattering particle velocity, $\underset{\sim}{v}_j = 0$,

$$\underset{\sim}{V}_{cm} = \frac{m_i}{m_i + m_j} \underset{\sim}{V}_i$$

and

$$\frac{1}{2}m_i v_i - \frac{1}{2}m_i v_i'^{\,2} = \frac{m_i^{\,2}}{m_i + m_j} v_i^{\,2} \left[1 - \frac{\underset{\sim}{v}_i \cdot \underset{\sim}{v}'_i}{v_i^{\,2}} \right] \qquad (11.25)$$

If the ion's velocity change during the collision is not large, then $\underset{\sim}{v}_i \cdot \underset{\sim}{v}'_i = v_i^{\,2} \cos \chi$, where as before χ is the scattering angle. Thus for small energy losses (11.25) becomes

$$\frac{1}{2}m_i v_i^{\,2} - \frac{1}{2}m_i v_i'^{\,2} \simeq \frac{m_i^{\,2}}{m_i + m_j} v_i^{\,2} (1 - \cos \chi),$$

or in terms of the incident ion energy, $U_i = m_i v_i^{\,2}/2$,

$$\frac{\Delta U_i}{U_i} \simeq \frac{2m_i}{m_i + m_j} (1 - \cos \chi) \qquad (11.26)$$

If the energy loss is averaged over χ,

$$\xi = \overline{\frac{\Delta U_i}{U_i}} = \frac{2m_i}{m_i + m_j} \qquad (11.27)$$

results. Differentiating (11.26) logarithmically gives

$$\frac{dv'_i}{v'_i} = \frac{dv_i}{v_i}$$

so

$$d\underset{\sim}{V}_{v_i} = 2\pi v_i^{\,2} \sin \theta \, d\theta \, dv_i$$

$$= 2\pi v_i^{\,2} \sin \theta \, d\theta \left(\frac{dv_i}{dv'_i}\right) dv'_i \qquad (11.28)$$

$$= \frac{v_i^{\,3}}{v_i'^{\,3}} d\underset{\sim}{V}_{v'_i}$$

These are the relations between v_i and v'_i, and dV_{v_i} and $dV_{v'_i}$, necessary to modify (11.24) for energy loss.

Write the integrand of (11.24) in the form

$$\left[f_i' v_i' I_{ij}' \frac{dV_{v_i'}}{dV_{v_i}} - f_i v_i I_{ij} \right] dV_{v_i}$$

$$= \left[\frac{f_i' v_i'^4 I_{ij}' - f_i v_i^4 I_{ij}}{v_i^3} \right] dV_{v_i}$$

$$\simeq \frac{v_i^2 - v_i'^2}{v_i^3} \frac{\partial (v_i^4 f_i I_{ij})}{\partial (v_i^2)} dV_{v_i} + v_i (f_i' - f_i) I_{ij} dV_{v_i}$$

using (11.28) and the Taylor series expansion of $v_i^4 I_{ij} f_i$. From (11.26),

$$\frac{v_i^2 - v_i'^2}{v_i^2} = \xi (1 - \cos \chi)$$

Thus the collision term (11.24) becomes

$$\frac{\delta f_i}{\delta t} = \sum_j n_j \left\{ \int \left[f_i(x; v_i, \theta', \phi'; t) - f_i(x; v_i, \theta, \phi; t) \right] v_i I_{ij} d\Omega + \frac{1}{v_i} \cdot \frac{\partial}{\partial (v_i^2)} \left[v_i^3 f_i \xi \int I_{ij} v_i (1 - \cos \chi) d\Omega \right] \right\} \quad (11.29)$$

If the expansion of f_i in spherical harmonics is now substituted into equation 11.29, the first term of the resultant expression has been previously obtained (equation 11.22). The second correction term is seen to contribute a scalar portion to the resulting collision term for the f_{i0} term, and this is the only one that will be retained. Thus in the spirit of the original collision term, I write

$$\frac{\delta f_i}{\delta t} = -\left[\nu_1 f_1 \cdot \frac{\mathbf{v}}{v} + .. \right] + \frac{1}{v_i} \frac{\partial}{\partial (v_i^2)} \left[v_i^3 f_{i0} \nu_1 \xi \right] \quad (11..0)$$

where ν_1 was defined in (11.21)., as the collision term corrected for an ion's energy loss during an encounter.

To correct the collision terms for the neglected energy of the scattering centers (i.e., their thermal energy) an argument of Davydov (1935) is employed. Assume that the scatterers have a Maxwellian distribution of velocities at a temperature $T = T_s$. By attaching a coordinate system to this scattering center, it is seen that the result of the inclusion of the scattering particle's energy is a modification of the ionic pre-collisional velocity:

$$\tilde{v}_i(T = T_s) \simeq v_i(T = 0) + 0(v_j)$$

Also then,

$$\tilde{f}_i(T = T_s) \simeq f_i(T = 0) + 0(v_j)\frac{\partial f_i}{\partial v_i} \tag{11.31}$$

If there were no external forces, $\underset{\sim}{F}_i$, and the ions were in equilibrium with the scattering particles, then the ionic distribution function would be a Maxwellian:

$$f_i = n_i\left(\frac{m_i}{2\pi kT_s}\right)^{3/2} \exp\left(-\frac{m_i v_i^2}{2kT_s}\right) \tag{11.32}$$

where k is Boltzmann's constant. Thus in this situation the left hand side of Boltzmann's equation is identically zero,and from the collision term the conditions

$$\underset{\sim}{\tilde{f}}_{i1} \equiv 0, \quad \underset{\approx}{\tilde{f}}_{i2} \equiv 0, \; \ldots$$

and

$$\frac{1}{v_i}\frac{\partial}{\partial(v_i^2)}\left[v_i^3 f_{i0} \nu_1 \xi\right] = 0 \tag{11.33}$$

result. Equation (11.33) implies $\tilde{f}_{i0} = 0$ and from (11.31) and (11.32) gives

$$0 = f_i + 0(v_j)\frac{\partial f_i}{\partial v_i} = f_i\left[1 - 0(v_j)\frac{mv_i}{kT_s}\right]$$

or

$$0(v_j) = \frac{kT_s}{mv_i}$$

Thus the collision term (11.30), corrected for the thermal motion of the scatterers, reads

$$\frac{\delta f_i}{\delta t} = -(\nu_1 \underset{\sim}{f}_{i1} + \nu_2 \underset{\approx}{f}_{i2} + \ldots) + \frac{1}{v_i}\frac{\partial}{\partial(v_i^2}\left[v_i^3 \nu_1 \xi (f_{i0} + \frac{kT_s}{mv_i}\frac{\partial f_{i0}}{\partial v_i})\right]$$

Note that this assumes the form obtained with the assumption of zero scatterer thermal energy when $T_s = 0$.

The collision term (11.30), corrected for scatterer termal motion is qualitatively identical with the collision term obtained by Rice and Gray (1965) for transport in liquids. The first term on the right hand side corresponds to a term they obtain for short range repulsive interactions. The second term corresponds to a term to describe the effects of long range interactions, essentially a friction drag term.

4. Decomposition of the Boltzmann Equation to a Set of Equations. Collecting the results of the two previous sections, Boltzmann's equation takes the form

$$\begin{aligned}
&\frac{\partial f_{i0}}{\partial t} + \frac{\underset{\sim}{v_i}}{v_i} \bullet \frac{\partial \underset{\sim}{f_{il}}}{\partial t} + \ldots \\
&+ \underset{\sim}{v_i} \bullet \nabla_{\underset{\sim}{x}} f_{i0} + \frac{\underset{\sim}{v_i}\underset{\sim}{v_i}}{v_i} : \nabla_{\underset{\sim}{x}} \underset{\sim}{f_{il}} + \ldots \\
&+ \frac{\underset{\sim}{F_i} \bullet \underset{\sim}{f_{il}}}{m_i v_i} + \frac{\underset{\sim}{F_i} \bullet \underset{\sim}{v_i}}{m_i v_i} \frac{\partial f_{i0}}{\partial v_i} + \frac{\underset{\sim}{F_i}}{m_i} \frac{\partial}{\partial v_i}\left(\frac{\underset{\sim}{f_{il}}}{v_i}\right) : \frac{\underset{\sim}{v_i}\underset{\sim}{v_i}}{v_i^2} + \ldots \\
&= -\nu_1 \underset{\sim}{f_{il}} \bullet \frac{\underset{\sim}{v_i}}{v_i} + \frac{1}{2v_i^2} \frac{\partial}{\partial v_i}\left[v_i^3 \nu_1 \xi \left(f_{i0} + \frac{kT_s}{mv_i} \frac{\partial f_{i0}}{\partial v_i}\right)\right]
\end{aligned} \qquad (11.34)$$

where terms f_{ij} for $j \geq 2$ have not been written down. It was stated previously that one of the reasons for expanding the distribution function in a series of orthogonal functions is to make use of the orthogonality properties in obtaining a chain of coupled equations from the Boltzmann equation. This procedure follows.

If equation (11.34) is multiplied by $R_{000} = 1$, and integrated over $d\Omega = \sin\theta \, d\theta d\phi$, the second, third, and sixth terms on the left hand side vanish as does the first term on the right hand side. The remaining terms give the equation

$$\begin{aligned}
&\frac{\partial f_{i0}}{\partial t} + \frac{1}{3} v_i \nabla_{\underset{\sim}{x}} \cdot \underset{\sim}{f_{il}} + \frac{\underset{\sim}{F_i} \bullet \underset{\sim}{f_{il}}}{m_i v_i} + \frac{v_i \underset{\sim}{F_i}}{3m_i} \bullet \frac{\partial}{\partial v_i}\left(\frac{\underset{\sim}{f_{il}}}{v_i}\right) \\
&= \frac{1}{2v_i^2} \frac{\partial}{\partial v_i}\left[v_i^3 \nu_1 \xi \left(f_{i0} + \frac{kT_s}{m_i v_i} \frac{\partial f_{i0}}{\partial v_i}\right)\right],
\end{aligned}$$

or upon combining the third and fourth terms,

$$\begin{aligned}
&\frac{\partial f_{i0}}{\partial t} + \frac{1}{3} v_i \nabla_{\underset{\sim}{x}} \cdot \underset{\sim}{f_{il}} + \frac{1}{3v_i^2} \frac{\partial}{\partial v_i}\left(v_i^2 \frac{\underset{\sim}{F_i} \bullet \underset{\sim}{f_{il}}}{m_i}\right) \\
&= \frac{1}{2v_i^2} \frac{\partial}{\partial v_i}\left[v_i^3 \nu_1 \xi \left(f_{i0} + \frac{kT_s}{m_i v_i} \frac{\partial f_{i0}}{\partial v_i}\right)\right]
\end{aligned} \qquad (11.35)$$

In a similar manner, if (11.34) is multiplied successively by R_{100}, R_{110}, and R_{111} and integrated each time over $d\Omega$, three equations result which, with the definition of equation 11.17, may be combined as one vector equation

$$\frac{\partial \underset{\sim}{f}_{i1}}{\partial t} + v_i \underset{\sim}{\nabla}_x f_{i0} + \frac{1}{m_i} \underset{\sim}{F}_i \frac{\partial f_{i0}}{\partial v_i} = -\nu_1 \underset{\sim}{f}_{i1} \qquad (11.36)$$

It is now assumed that the forces, $\underset{\sim}{F}_i$, which have been hitherto unspecified except for their velocity independence, are due solely to an external electric field $\underset{\sim}{E}$ Thus

$$\underset{\sim}{F}_i = q_i \underset{\sim}{E} = eZ_i \underset{\sim}{E}$$

where z_i is the signed valence of the i th type of ion. If $\underset{\sim}{a}_i$ is the acceleration that the ion experiences due to the field,

$$\underset{\sim}{a}_i = \frac{1}{m_i} \underset{\sim}{F}_i = \frac{q_i}{m_i} \underset{\sim}{E}$$

and therefore equations 11.35 and 11.36 may be rewritten as

$$\frac{\partial f_{i0}}{\partial t} + \frac{1}{3} v_i \underset{\sim}{\nabla}_x \cdot \underset{\sim}{f}_{i1} + \frac{g_i}{3 v_i^2} \cdot \frac{\partial}{\partial v_i} (v_i^2 \underset{\sim}{f}_{i1})$$
$$= \frac{1}{2 v_i^2} \frac{\partial}{\partial v_i} \left[v_i^3 \nu_1 \xi \left(f_{i0} + \frac{kT_s}{m_i v_i} \frac{\partial f_{i0}}{\partial v_i} \right) \right] \qquad (11.37)$$

and

$$\frac{\partial \underset{\sim}{f}_{i1}}{\partial t} + v_i \underset{\sim}{\nabla}_x f_{i0} + \underset{\sim}{a}_i \frac{\partial f_{i0}}{\partial v_i} = -\nu_1 \underset{\sim}{f}_{i1} \qquad (11.38)$$

G. Dependence of the Collision Frequency, ν_1, on v_i.

1. General Considerations. Equations 11.37 and 11.38 contain ν_1, defined by (11.21) as

$$\nu_1(v_i) = \sum_j n_j v_i \int_0^{\pi} (1-\cos \chi) I_{ij}(v_i, 2) d\Omega,$$

which depends on ionic velocity, v, and which must be determined before the equations can be solved. In the expression for ν_1, $d\Omega = 2\pi \sin \chi d\chi$ while from (11.19)

$$I(v,\chi) = \frac{b}{\sin \chi} \quad \frac{db}{d\chi}$$

Thus in order to evaluate the integral for ν_1, a relation between b and χ must be determined.

From the discussion of a binary collision in section D it is clear that for a spherically symmetric potential function the trajectory of the equivalent particle, of reduced mass μ, is symmetric about the midpoint of the collision, (r_m, θ_m). Thus the scattering angle χ is related to θ_m by

$$\chi = \pi - 2\theta_m \tag{11.39}$$

In addition

$$\theta_m = \int_0^{\theta_m} d\theta = \int_\infty^{r_m} \left(\frac{d\theta}{dr}\right) dr = \int_\infty^{r_m} \left[\frac{d\theta}{dt} \Big/ \frac{dr}{dt}\right] dr \tag{11.40}$$

If the relative pre-collisional velocity, g(t), of Figure 11.2 is written in polar coordinates (r, θ)

$$\underset{\sim}{g}(\dot{r}) = \hat{\underset{\sim}{r}}\frac{dr}{dt} + \hat{\underset{\sim}{\theta}}\, r\frac{d\theta}{dt},$$

where "$\wedge$" indicates a unit vector, then using the fact that angular momentum is conserved

$$\mu bg = |\mu \underset{\sim}{r} \times \underset{\sim}{g}(t)|$$

$$= \mu r^2 \frac{d\theta}{dt},$$

where μbg is the precollisional angular momentum. Thus

$$\frac{d\theta}{dt} = \frac{bg}{r^2} \tag{11.41}$$

From the conservation of energy for the equivalent particle

$$\tfrac{1}{2}\mu g^2 = \tfrac{1}{2}\mu [\,g(t)]^2 + \varphi(r),$$

where $\mu g^2/2$ is the pre-collisional energy, the relation

$$\begin{aligned} g^2 &= \underset{\sim}{g}(t) \cdot \underset{\sim}{g}(t) + \frac{2}{\mu}\varphi(r) \\ &= \left(\frac{dr}{dt}\right)^2 + \left(\frac{d\theta}{dt}\right)^2 r^2 + \frac{2}{\mu}\varphi(r) \end{aligned} \tag{11.42}$$

follows. Combining equations 11.41 and 11.42, and solving for $\dot{r}$ gives

$$\frac{dr}{dt} = -g\left[1 - \frac{b^2}{r^2} - \frac{2\varphi(r)}{\mu g^2}\right]^{1/2} \tag{11.43}$$

Finally, using equations 11.39 to 11.41 and (11.43)

$$\chi = \pi - 2b\int_{r_m}^{\infty} r^{-2}\left[1 - \frac{b^2}{r^2} - \frac{2\varphi}{\mu g^2}\right]^{-\frac{1}{2}} dr \tag{11.44}$$

To obtain r_m for the lower integration limit in (11.44) it is sufficient to invoke the condition

$$\left(\frac{dr}{d\theta}\right)_{(r_m,\theta_m)} = \left(\frac{\dot{r}}{\dot{\theta}}\right)_{(r_m,\theta_m)} = 0$$

With $\dot{\Theta}$ always finite this implies $\dot{r}(r_m, \Theta_m) \equiv 0$ so (11.43) gives

$$1 = \frac{b^2}{r_m{}^2} + \frac{2\varphi(r_m)}{\mu g^2} \tag{11.45}$$

an implicit relation for r_m. The expressions (11.9), (11.44) and (11.45) and any intermolecular potential function permit us to write $\nu_1(v)$ explicitly.

Assume that the force between the ion i and scatterer j during a collision is given by

$$|\underset{\sim}{F}| = -\frac{k_{ij}}{r^{\alpha}}$$

The potential is thus

$$\varphi(r) = \frac{-k_{ij}}{(\alpha-1)r^{\alpha-1}}$$

Define two dimensionless numbers

$$u = \frac{b}{r} \text{ and } u_0 = b\left\{-\frac{\mu v_i{}^2}{k_{ij}}\right\}^{\frac{1}{\alpha-1}}$$

With these new definitions, the scattering angle may be rewritten as

$$\chi = \pi - 2\int_0^{u_\alpha} \frac{du}{\left[1 - u^2 - \frac{2}{\alpha-1}\left(\frac{u}{u_0}\right)^{\alpha-1}\right]^{1/2}}$$

where $u_\alpha = b/r_\alpha$. The expression for the collision frequency ν_1 is

$$\begin{aligned}\nu_1 &= 2\pi \sum_j n_j v_i \int_0^{\pi} (1-\cos\chi)\left(\frac{-k_{ij}}{\mu v_i}\right)^{\frac{2}{\alpha-1}} u_0 \left|\frac{du_0}{d\chi}\right| d\chi \\ &= 2\pi \sum_j n_j v_i \left(\frac{k_{ij}}{\mu v_i{}^2}\right)^{\frac{2}{\alpha-1}} A_{ij}(\alpha)\end{aligned} \tag{11.46}$$

where

$$A_{ij}(\alpha) = \int_0^{\pi} (-1)^{\frac{2}{\alpha-1}} (1-\cos\chi) u_0 \left|\frac{du_0}{dx}\right| dx$$

is a dimensionless number depending only on m. The velocity dependence of (11.46) is

$$\nu_1 = \beta_{ij} v_i{}^p \tag{11.47}$$

where

$$\beta_{ij} = 2\pi \sum_j n_j \left(\frac{k_{ij}}{\mu}\right)^{\frac{2}{\alpha-1}} A_{ij} \tag{11.47a}$$

TABLE 11.1

Interaction Type	$\varphi(r)$	α	p	$\nu_1(v_i)/\beta_{ij}$	β_{ij}
Ion-charged particle	$\frac{q_i q_j}{4\pi\varepsilon_o r}$	2	-3	v_i^{-3}	$2\pi \sum_j \left(\frac{q_i q_j}{4\pi\varepsilon_o \mu}\right)^2 n_j A_{ij}(2)$
Ion-permanent dipole	$-\frac{q_i \bar{\mu}_j \cos\delta}{4\pi\varepsilon_o r^2}$	3	-1	v_i^{-1}	$2\pi \sum_j n_j \left(\frac{q_i \bar{\mu}_j \cos\delta}{2\pi\varepsilon_o \mu}\right) A_{ij}(3)$
Ion-induced dipole	$-\frac{\alpha_j q_i^2}{8\pi\varepsilon_o r^4}$	5	0	1	$2\pi \sum_j n_j q_i \left(\frac{\alpha_j}{2\pi\varepsilon_o \mu}\right)^{1/2} A_{ij}(5)$
Induced dipole-induced dipole	$-\frac{3\alpha_i \alpha_j}{8\pi\varepsilon_o r^6} \cdot \frac{I_i I_j}{(I_i+I_j)}$	7	1/3	$v_i^{1/3}$	$2\pi \sum_j n_j \left(\frac{9\alpha_i \alpha_j}{4\pi\varepsilon_o} \cdot \frac{I_i I_j}{I_i+I_j}\right)^{1/3} A_{ij}(7)$
Hard sphere	$\varphi(r) = \begin{cases} 0 & r > r_i + r_j \\ \infty & r \leq r_i + r_j \end{cases}$	∞	1	v_i	$2\pi \sum_j n_j (r_i+r_j)^2 A_{ij}(\infty)$

Table 11.1. The central ion-membrane molecule interactions considered and the corresponding collision frequency. q = charge, $\bar{\mu}\cos\delta$ = dipole moment perpendicular to membrane, α = polarizability, I = second ionization potential, subscripts i and j refer to ion and scatterer respectively. $r_{i(j)}$ = ion (scatterer) radius.

and

$$p = (\alpha - 5)/(\alpha - 1)$$

In Table 11.1 I have listed the five primary interactions between ion and membrane molecule that will be extensively used later. The velocity dependence of the collision frequency corresponding to each is also listed as well as the dependence of β_{ij} on various ion and scatterer parameters. The values of the numbers $A_{ij}(\alpha)$ are given in Table 11.2.

TABLE 11.2

α	$A_{ij}(\alpha)$
2	
3	
5	0.422
7	0.385
∞	0.5

Table 11.2. Values of the parameter $A_{ij}(\alpha)$ appearing in equation 11.47a as a function of α.

H. Expressions for Macroscopic Quantities from the Expanded Distribution Function.

The early part of this chapter shows how various macroscopic quantities are formally related to the distribution function. With the expanded distribution function the various orthogonality relations change the form of the previous expressions slightly.

The original equation for the number density is

$$n_i = \int f_i dV_{\underset{\sim}{v_i}}$$

$$= \int f_i v_i^2 dv_i \sin\theta \, d\theta d\phi$$

If the spherical harmonic expansion of f_i is substituted into this expression, and the integrations over θ and ϕ are carried out, all that remains is

$$n_i = 4\pi \int f_{i0} v_i^2 dv_i \qquad (11.48)$$

All of the other terms integrate to zero.

The current density due to a charged species was shown to be

$$\underset{\sim}{j}_i = q_i \int f_i \underset{\sim}{v}_i dV_{\underset{\sim}{v}_i}$$

$$= q_i \int f_i \underset{\sim}{v}_i v_i^2 dv_i \sin\Theta\, d\Theta d\phi$$

If $\underset{\sim}{v}_i$ is written as

$$\underset{\sim}{v}_i = v_i \left[\underset{\sim}{i}_x \sin\Theta \cos\phi + \underset{\sim}{i}_y \sin\Theta \sin\phi + i_z \cos\Theta \right]$$

this and the expansion for f_i are substituted into the current density equation, and all of the integrations over Θ and ϕ carried out the result is

$$\underset{\sim}{j}_i = \frac{4nq_i}{3} \int \left[\underset{\sim}{i}_x f_{i110} + \underset{\sim}{i}_y f_{i111} + \underset{\sim}{i}_z f_{i100} \right] v_i^3 dv_i$$

or

$$\underset{\sim}{j}_i = \frac{4nq_i}{3} \int v_i^3 \underset{\sim}{f}_{i1} dv_i \qquad (11.49)$$

Thus the current density depends directly on $\underset{\sim}{f}_{i1}$ only, though in general this will be a function of f_{i0}.

I. Summary.

If $\underset{\sim}{x}$ and $\underset{\sim}{v}$ (a "tilda" under a quantity denotes a vector) respectively denote vector position and velocity of a particle of the i th ionic species, then the DISTRIBUTION FUNCTION $f(\underset{\sim}{x},\underset{\sim}{v},t)$ for this species is defined as the number of ions of the i th species which, at a time t, are in a spatial volume element ($d^3\underset{\sim}{x}$, cm^3) about $\underset{\sim}{x}$(cm) and a velocity volume element ($d^3\underset{\sim}{v}$, cm^3/sec^3) about $\underset{\sim}{v}$(cm/sec).

The distribution function is of considerable use by virtue of its relation to measurable macroscopic variables. For example, the number density $n(\underset{\sim}{x},t)$ number/cm^3 of particles at a particular $(\underset{\sim}{x},t)$ may be obtained by summing (integrating) the number of particles in each velocity range:

$$n(\underset{\sim}{x},t) = \int d^3\underset{\sim}{v} f(\underset{\sim}{x},\underset{\sim}{v},t)$$

In a similar manner, the number flux [$\underset{\sim}{N}(\underset{\sim}{x},t)$ number/cm^2-sec] of particles crossing a unit surface area is

$$\underset{\sim}{N}(\underset{\sim}{x},t) = \int d^3\underset{\sim}{v}\, \underset{\sim}{v} f(\underset{\sim}{x},\underset{\sim}{v},t)$$

If the particles have a charge $q=ze$ then the current density, $\underset{\sim}{j}$ (A/cm^2) carried by this flux is $\underset{\sim}{j} = q\underset{\sim}{N}(\underset{\sim}{x},t)$. In general the distribution function will be a function of various external driving forces; e.g., gradients of concentration, temperature, or electrical potential.

The distribution function must be expressed as an explicit function of velocity so the integrations of the previous section may be carried out. With all of the assumptions listed in the introduction to Part III, except for number 3, the following equation of change for f, commonly known as the Boltzmann transport equation, results (Green, 1952):

$$\frac{\partial f}{\partial t} + \underset{\sim}{v}.\nabla_{\underset{\sim}{x}} f + (\underset{\sim}{F}/m).\nabla_{\underset{\sim}{v}} f = \iint d^3\underset{\sim}{v}_s d\Omega (f' f'_s - f f_s) g_{is} I_{is} \qquad (11.5) + (11.13)$$

In this equation m is the mass (gm/particle) of a particle of the i th ionic species, $\underset{\sim}{F}$ is an external force (dynes/particle) acting on the i th species, f_s is the distribution function for the scatterer molecules, and $g_{is} = |\underset{\sim}{v}_i - \underset{\sim}{v}_s|$ is the speed (cm/sec) of an ion relative to a scattering particle. I_{is} is a differential scattering cross section (cm^2), $d\Omega = 2\pi \sin\chi \, d\chi$ is a solid angle (steradians) and χ is the angle (radians) through which the ion is deflected during a collision. (See Goldstein, 1950, for a discussion of the mechanics of binary collisions). Primed distribution functions refer to a post-collisional state while unprimed ones are precollisional.

The Boltzmann equation is a conservation equation for $f(\underset{\sim}{x},\underset{\sim}{v},t)$ and some insight into the origin of the various terms is desirable. Typical conservation equations relate the time rate of change of a quantity to the divergence of its flux. The present case requires a relation between the time rate of change of $f(\underset{\sim}{x},\underset{\sim}{v},t)$ and the divergence of its flux in $\underset{\sim}{x}$ space, $\nabla_x \cdot [f(d\underset{\sim}{x}/dt)]$, the divergence of its flux in $\underset{\sim}{v}$ space, $\nabla_v \cdot [f(d\underset{\sim}{v}/dt)]$, and the rate of change of $f(\underset{\sim}{x},\underset{\sim}{v},t)$ due to collisions. The integral expression on the right hand side gives the rate of change of $f(\underset{\sim}{x},\underset{\sim}{v},t)$ due to collisions, $\underset{\sim}{v}\cdot\nabla_{\underset{\sim}{x}} f$ is due to flow in position space, and $\underset{\sim}{a}\cdot\nabla_{\underset{\sim}{v}} f$ arises because of flow in velocity space in the presence of a velocity independent acceleration $\underset{\sim}{a} = \underset{\sim}{F}/m$.

Solutions to the Boltzmann equation are usually based on an expansion of $f(\underset{\sim}{x},\underset{\sim}{v},t)$ in terms of some suitably small parameter. The form of the expansion used here is suggested by expressing the ion velocity $\underset{\sim}{v}$ in spherical coordinates (v,θ,ϕ), where θ and ϕ are now the variables during a collision, and expressing the distribution function as $f(\underset{\sim}{x},\underset{\sim}{v},t) = f(\underset{\sim}{x};v;\theta,\phi,t)$. This may be expanded in spherical harmonics. Such an expansion is equivalent (Johnson, 1960, 1966) to a tensor expansion

$$f = f_0(\underset{\sim}{x},\underset{\sim}{v},t) + (\underset{\sim}{v}/v)\cdot\underset{\sim}{f_1} + (\underset{\sim}{vv}/v^2):\underset{\approx}{f_2} + \ldots \qquad (11.16)$$

If the expansion (11.16) is substituted into the Boltzmann equation, and appropriate moments of the equation taken, equations for $f_0, \underset{\sim}{f_1}, \underset{\approx}{f_2}$ ... result. Morrone (1968) has shown that terms f_j in the expansion (11.16) are proportional to $(\xi)^{j/2}$, where ξ was defined earlier as the ion fractional energy loss during an ion-scatterer collision. Thus with assumption 3 that $\xi \ll 1$, terms f_j for $j > 1$ may be neglected in (11.16). Therefore, the expansion is written as $f = f_0(\underset{\sim}{x},\underset{\sim}{v},t) + (\underset{\sim}{v}/v)\cdot\underset{\sim}{f_1}(\underset{\sim}{x},\underset{\sim}{v},t)$.

If the acceleration a of the ions due to external forces is assumed to be in the $\underset{\sim}{x}$ direction, perpendicular to the membrane surface, and assumption 3 now holds, the equations

$$\frac{\partial f_0}{\partial t} + \frac{1}{3}v\frac{\partial f_1}{\partial x} + \frac{a}{3v^2}\frac{\partial (v^2 f_1)}{\partial v} = S(f_0) \qquad (11.35)$$

and

$$\frac{\partial f_1}{\partial t} + v\frac{\partial f_0}{\partial x} + a\frac{\partial f_0}{\partial v} = -\nu_1(v) f_1 \qquad (11.36)$$

result (Ginzburg and Gurvich, 1960). In (11.35)

$$S(f_0) = \frac{1}{2v^2}\frac{\partial}{\partial v}\left[v^3\nu_1\xi\left(f_0 + \frac{kT}{mv}\frac{\partial f_0}{\partial v}\right)\right]$$

The fractional energy loss $\xi = 2m/(m + m_s)$, where m_s is the scatterer mass; k and T are, respectively, Boltzmann's constant (erg/^{o}K) and the absolute temperature (^{o}K) of the system. The frequency of collisions of an ion of the i th species with the membrane scatterers, $\nu(v)$ (collisions/sec), is given by

$$\nu(v) = n_s \int_0^{\pi} d\Omega (1-\cos\chi) I_{is}$$

where n_s is the number density of scatterers. ν will, in general be a function of the ionic velocity v.

In order to obtain a solution to equations 11.35 and 11.36 for f_0 and f_1, and thus enable a calculation of current flow through the membrane model as a function of external driving forces, ν must be expressed as an explicit function of v. The form of this velocity dependence is determined by the nature of the intermolecular forces between ion and membrane molecule. For central conservative forces there is a simple power law relation between ν and v.

If, as assumption 2 requires, the force F_{is} between ion and scatterer is given by $F_{is} = -(K_{is}/r_{is}^{\alpha})$ where r_{is} is the separation between ion and scatterer, and α is a constant, then (Chapman and Cowling, 1939) the collision frequency is given by

$$\nu(v) = \beta_i v^p \tag{11.47}$$

where $p = (\alpha-5)/(\alpha-1)$ and

$$\beta_i = 2\pi n_s A(\alpha) \left[K_{is}(m+m_s)/mm_s \right]^{2/(\alpha-1)} \tag{11.47a}$$

is a constant involving ionic and scatterer parameters. $A(\alpha)$ is a pure number depending only on α.

Equations 11.35-36, and 11.47, complete the necessary set needed for a calculation of f_0 and f_1. In terms of f_0 and f_1, integral expressions for the number density and current density become

$$n(x,t) = 4\pi \int_0^{\infty} dv\, v^2 f_0 \tag{11.48}$$

and

$$j(x,t) = (4\pi q/3) \int_0^{\infty} dv\, v^3 f_1 \tag{11.49}$$

respectively.

CHAPTER 12. RELATIONSHIP BETWEEN THE MICROSCOPIC AND MACROSCOPIC FORMULATIONS OF ELECTRODIFFUSION THEORY.

In Chapter 6 I developed a macroscopic approach to trans-membrane ion movement in which physical intuition was employed to develop the various ionic conservation equations. The same basic physical model was developed in the previous chapter, but from a microscopic basis. It is the purpose of this chapter to examine the connection between the two approaches.

A. Conservation Equations.

First, I will be specifically concerned with deriving conservation of mass and energy equations from

$$\frac{\partial f_0}{\partial t} + \frac{v}{3}\nabla_{\underset{\sim}{x}} \cdot \underset{\sim}{f}_1 + \frac{1}{3v^2}\frac{\partial (v^2 \underset{\sim}{a} \cdot \underset{\sim}{f}_1)}{\partial v} = S(f_0) \tag{11.35}$$

and a conservation of momentum equation from

$$\frac{\partial \underset{\sim}{f}_1}{\partial t} + v\nabla_{\underset{\sim}{x}} f_0 + \underset{\sim}{a}\frac{\partial f_0}{\partial v} = -\nu_1 \underset{\sim}{f}_1 \tag{11.36}$$

where

$$S(f_0) = \frac{1}{2v^2}\frac{\partial}{\partial v}\left[\xi v^3 \nu_1 (f_0 + \frac{kT}{mv}\frac{\partial f_0}{\partial v})\right] \tag{12.1}$$

and the ionic species identifying subscript 'i' has been deleted. Before attempting this, it is well to note that the average value of any scalar function $G(v)$ of the velocity magnitude v is given by

$$\overline{G(v)} = \frac{\int G(v)\, f(\underset{\sim}{x},\underset{\sim}{v},t)\, d^3\underset{\sim}{v}}{\int f(\underset{\sim}{x},\underset{\sim}{v},t)\, d^3\underset{\sim}{v}}$$

$$= \frac{1}{n} \int G(v) f(\underset{\sim}{x}, \underset{\sim}{v}, t) d^3 v \qquad (12.2)$$

wherein n is the ionic number density. Using the expanded form of the distribution function, (12.2) may be immediately written as

$$\overline{G(v)} = \frac{1}{n} \int_0^\infty 4\pi v^2 G(v) f_0 dv \qquad (12.3)$$

In a like manner, the average of any vector function $\underset{\sim}{C} = C(v)\underset{\sim}{v}/v$ is given by

$$\overline{\underset{\sim}{C}} = \frac{1}{n} \int C(v) \frac{\underset{\sim}{v}}{v} f(\underset{\sim}{x}, \underset{\sim}{v}, t) d^3 v$$

$$= \frac{1}{n} \int_0^\infty \left(\frac{4\pi}{3}\right) v^2 C(v) \underset{\sim}{f_1} dv \qquad (12.4)$$

Thus from the above considerations if (11.35) is multiplied by $4\pi v^2 dv$ and integrated over the velocity range the first and second terms give $(\partial n/\partial t)$ and $\nabla_{\underset{\sim}{x}} \cdot (n\underset{\sim}{\bar{v}})$ respectively. Therefore

$$\frac{\partial n}{\partial t} + \nabla_{\underset{\sim}{x}} \cdot (n\underset{\sim}{\bar{v}}) = 4\pi \int_0^\infty \frac{\partial}{\partial v}\left[\frac{\xi v^3 \nu_1}{2}\left(f_0 + \frac{kT}{mv}\frac{\partial f_0}{\partial v}\right) - \frac{v^2 \underset{\sim}{a} \cdot \underset{\sim}{f_1}}{3}\right] dv$$

$$= 4\pi \left[\frac{\xi v^3 \nu_1}{2}\left(f_0 + \frac{kT}{mv}\frac{\partial f_0}{\partial v}\right) - \frac{v^2 \underset{\sim}{a} \cdot \underset{\sim}{f_1}}{3}\right]_0^\infty$$

Now I assume that f_0 and f_1 tend to zero as v goes to infinity much faster than any of the velocity dependent quantities so the first equation, a conservation of mass (ion) equation is

$$\frac{\partial n}{\partial t} + \nabla_{\underset{\sim}{x}} \cdot (n\underset{\sim}{\bar{v}}) = 0 \qquad (12.5)$$

Writing (12.5) in one dimension, x, perpendicular to the membrane surface and realizing that $\bar{v}$ is just the directed ionic velocity in the x direction results in a recovery of equation 6.9

$$\frac{\partial n}{\partial t} + \frac{\partial (n\bar{v}_d)}{\partial x} = 0 \qquad (6.9)$$

If equation 11.35 is multiplied by $(mv^2/2)(4\pi v^2 dv)$, where $u = mv^2/2$ is the ionic energy, and again integrated over all v then from (12.3) and (12.4)

$$\frac{\partial(nu)}{\partial t} + \nabla_{\underset{\sim}{x}} \cdot (nu\underset{\sim}{\bar{v}}) + \int_0^\infty \frac{4\pi u}{3} \frac{\partial(v^2 \underset{\sim}{a} \cdot \underset{\sim}{f}_1)}{\partial v} dv = \int_0^\infty u\, S(f_0) 4\pi v^2 dv \qquad (12.6)$$

results. Now

$$\int_0^\infty \frac{4\pi u}{3} \frac{\partial(v^2 \underset{\sim}{a} \cdot \underset{\sim}{f}_1)}{\partial v} dv = \frac{4\pi}{3} \int_0^\infty u\, d(v^2 \underset{\sim}{a} \cdot \underset{\sim}{f}_1)$$

$$= -\frac{4\pi m}{3} \int v^3 \underset{\sim}{a} \cdot \underset{\sim}{f}_1 dv$$

$$= -nm\underset{\sim}{a} \cdot \underset{\sim}{\bar{v}}$$

and

$$\int_0^\infty u\, S(f_0) 4\pi v^2 dv = \pi m \int_0^\infty v^2 \frac{\partial}{\partial v} \left[\xi v^3 \nu_1 (f_0 + \frac{kT}{mv} \frac{\partial f_0}{\partial v}) \right] dv$$

$$= -2\pi m \int_0^\infty \xi v^4 \nu_1 (f_0 + \frac{kT}{mv} \frac{\partial f_0}{\partial v}) dv$$

$$= -\xi n \overline{\nu_1 u} - 2\pi kT \int_0^\infty \xi v^3 \nu_1 \frac{\partial f_0}{\partial v} dv$$

$$= -\xi n \overline{\nu_1 u} + \xi n u_s \overline{\nu_1}$$

wherein $u_s = 3kT/2$ is the scatterer energy. Thus (12.6) becomes

$$\frac{\partial(nu)}{\partial t} + \nabla_{\underset{\sim}{x}} \cdot (nu\underset{\sim}{\bar{v}}) - nm\underset{\sim}{a} \cdot \underset{\sim}{\bar{v}} = -\xi n(\overline{\nu_1 u} - \overline{\nu_1} u_s) \qquad (12.7)$$

Again writing this equation for a one dimensional situation, and taking $\underset{\sim}{a} = q\underset{\sim}{E}/m$ where $\underset{\sim}{E}$ is the electric field strength,

$$\frac{\partial(nu)}{\partial t} = nq\underset{\sim}{E}\bar{v} - \frac{\partial(nu\bar{v})}{\partial x} - \xi n(\overline{\nu_1 u} - \overline{\nu_1} u_s) \qquad (12.8)$$

results. Equation 12.8 is identical with the conservation equation, (6.6), if $(\overline{\nu_1 u} - \overline{\nu_1} u_s) = \overline{\nu_1}(u - u_s)$.

Finally, to derive an equation for the conservation of momentum from the microscopic equations multiply equation 11.36 by $(4\pi v^2/3)$ dv(mv) and integrate over all velocities to give

$$\frac{\partial(nm\underset{\sim}{\bar{v}})}{\partial t} + \nabla_{\underset{\sim}{x}}(n\bar{u}) - \underset{\sim}{a}nm = -\overline{\nu_1 u}mn \qquad (12.9)$$

Again in one dimension, (12.9) becomes

$$\frac{\partial (nm\bar{v})}{\partial t} = qEn - mn\overline{\nu_1 v} - \frac{\partial (n\bar{u})}{\partial x} \tag{12.10}$$

B. Relation to Electrodiffusion.

In this section equation 11.36 is used to relate ionic current density to a concentration gradient and external electric field. I first treat a constant (in time) electric field, and then consider the case of a step change in the field, to obtain equations that bear a superficial resemblance to the electrodiffusion equation. However, the mobility and diffusion coefficient are potentially functions of electric field strength, distance through the membrane, and time. Conditions are explored under which the electrodiffusion equation (with μ and D obeying the Einstein relation) is obtained.

In a truly time independent situation, $\dot{f}_1 = 0$ and from (11.36)

$$f_1 = -\frac{v}{\nu}\frac{\partial f_0}{\partial x} - \frac{a}{\nu}\frac{\partial f_0}{\partial v} \tag{12.11}$$

Thus, the current density is given by

$$j = (4\pi q/3)\left[-\frac{qE}{m}\int_0^\infty \frac{dvv^3}{\nu}\frac{\partial f_0}{\partial v} - \frac{\partial}{\partial x}\int_0^\infty \frac{dvv^4 f_0}{\nu}\right] \tag{12.12}$$

from (11.49). Equation 12.12 may also be written in the form

$$j = q^2 n\mu E - q\frac{\partial (nD)}{\partial x} \tag{12.13}$$

where the mobility, μ, is defined by

$$\mu(x,E) = \frac{\int_0^\infty \frac{dvv^3}{\nu}\left(\frac{\partial f_0}{\partial v}\right)}{3m\int_0^\infty dv\, v^2 f_0} \tag{12.14}$$

and the diffusion coefficient D, is

$$D(x,E) = \frac{\int_0^\infty \frac{dvv^4 f_0}{\nu}}{3\int_0^\infty dv\, v^2 f_0} \tag{12.15}$$

In (12.14) and (12.15) ionic and membrane (scatterer) molecular characteristics enter through both ν and f_0. If f_0 is to be determined so D and μ may be computed from equations 12.14 and 12.15, the following considerations are important. In a steady state $\dot{f}_0 = 0$, and thus $\dot{n} = 0$ by (11.48). From the continuity equation therefore, $(\partial j/\partial x) = 0$ and thus $(\partial f_1/\partial x) = 0$ by (11.49). The foregoing allows equations 11.35, 12.1, and 12.11 to be combined into a single equation for f_0:

$$(1 + \frac{2ma^2}{3kT\nu^2\xi})\frac{\partial f_0}{\partial v} + \frac{2ma^2v}{3kT\nu^2\xi}\frac{\partial f_0}{\partial x} + \frac{mv}{kT}f_0 = 0 \qquad (12.16)$$

Solution of this equation gives an explicit form for f_0 to be substituted into (12.14) and (12.15), and the indicated integrations may then be carried out. Solutions of (12.16) are considered in Chapters 13 and 14.

Equation 12.13 is similar to the electrodiffusion equation 6.19. However, it is not necessarily true that the Einstein relation is valid (see Chapter 6). Generally, $D = D(x)$ and $\mu = \mu(x)$ unless (1) f_0 is independent of x, or (2) the spatial dependence of f_0 is such that it may be factored into the product of spatial and velocity dependent terms. From (12.14) and (12.15) the Einstein relation, $D = \mu kT$, holds only when

$$\int_0^\infty \frac{dvv^3}{\nu}\left[\frac{\partial f_0}{\partial v} + \frac{mv}{kT}f_0\right] = 0 \qquad (12.17)$$

Thus, Einstein's relation is valid only when f_0 is of Maxwellian form, i.e.when $f_0 \sim \exp(-mv^2/2kT)$. Also, it is clear that when f_0 is Maxwellian neither the mobility nor the diffusion coefficient are dependent on x. Considerations of Chapters 13 and 14 indicate that in a steady state f_0 will be Maxwellian only for electric field strengths so small that the energy of an ion in the field is much less than its thermal energy. For appreciable field strengths f_0 will not be Maxwellian and will, in fact be a complicated function of E.

In a time dependent situation an equation similar to (12.13) is found. If a step field $E = E_0u(t)$ is applied, where

$$u(t) = \begin{cases} 0 & t < 0 \\ 1/2 & t = 0 \\ 1 & t > 0 \end{cases}$$

then the solution of (11.36) is

$$f_1 = -\frac{qE_0}{m}\int_0^t du e^{-\nu u}\frac{\partial f_0(x,v,t-u)}{\partial v} - v\frac{\partial}{\partial x}\int_0^t du e^{-\nu u} f_0(x,v,t-u) \tag{12.18}$$

Thus the current density is given by

$$j = (4\pi q/3)\left[-\frac{qE_0}{m}\int_0^\infty dv v^3 \int_0^t du e^{-\nu u}\frac{\partial f_0(x,v,t-u)}{\partial v} - \frac{\partial}{\partial x}\int_0^\infty dv v^4 \int_0^t du e^{-\nu u} f_0(x,v,t-u)\right] \tag{12.19}$$

As before, a mobility

$$\mu(x,t,E) = \frac{\int_0^\infty dv v^3 \int_0^t du e^{-\nu u}\left[-\frac{\partial f_0(x,v,t-u)}{\partial v}\right]}{3m\int_0^\infty dv v^2 f_0(x,v,t)} \tag{12.20}$$

and diffusion coefficient

$$D(x,t,E) = \frac{\int_0^\infty dv v^4 \int_0^t du e^{-\nu u} f_0(x,v,t-u)}{3\int_0^\infty dv v^2 f_0(x,v,t)} \tag{12.21}$$

may be defined and an equation like (12.13) written. As $t \to \infty$ equation 12.20 reduces to (12.14), while (12.21) reduces to (12.15). These general time dependent expressions for μ and D give the same restrictions as found before for the Einstein relation to be valid.

An equation in $f_0(x,v,t)$, analogous to the one presented in the steady state case, is obtained by combining equations 11.35, 12.1, and 12.18.

Before finishing this chapter I would like to point out that use of the dimensionless variables defined in Chapter 6 allows equations

11.35 and 12.1 to be written in the form

$$\frac{t_c}{\langle t_D \rangle} \frac{\partial f_0}{\partial T} + \frac{t_c}{t_D} V_d \frac{\partial f_1}{\partial X} + \frac{Z\bar{E}}{V_d^2} \frac{\partial (V_d^2 f_1)}{V_d}$$
$$= \frac{1}{2 V_d^2} \frac{\partial}{\partial V_d} \left[\xi \, V_d^{p+3} \left(f_0 + \frac{t_D}{t_c V_d} \frac{\partial f_0}{\partial V_d} \right) \right] \tag{12.22}$$

while (11.36) becomes

$$\frac{t_c}{\langle t_D \rangle} \frac{\partial f_1}{\partial T} + \frac{t_c}{t_D} V_d \frac{\partial f_0}{\partial X} + Z\bar{E} \frac{\partial f_0}{\partial V_d} = - V_d^p f_1 \tag{12.23}$$

A second, dimensionless, formulation of (11.35) and (12.1), and (11.36), which avoids the presence of the ratios $(t_c/\langle t_D \rangle)$ and (t_c/t_D) is possible if I define

$$V = v/v_T \qquad I = j/ev_T\bar{n}\sqrt{\xi}$$
$$\bar{X} = x\nu_0/v_T \qquad \tilde{E} = eE/m\nu_0 v_T \sqrt{\xi}$$
$$\bar{T} = \nu_0 t \qquad mv_T^2 = 3kT$$

instead of the corresponding variables of Chapter 6. With this set (11.35) and (12.1) become

$$\frac{\partial f_0}{\partial \bar{T}} + \frac{V}{3} \frac{\partial f_1}{\partial \bar{X}} + \frac{Z\tilde{E}}{3V^2} \frac{\partial (V^2 f_1)}{\partial V} = \frac{1}{2V^2} \frac{\partial}{\partial V} \left[V^{p+3} \left(f_0 + \frac{1}{3V} \frac{\partial f_0}{\partial V} \right) \right] \tag{12.24}$$

while equation 11.36 takes the form

$$\frac{\partial f_1}{\partial \bar{T}} + V \frac{\partial f_0}{\partial \bar{X}} + Z\tilde{E} \frac{\partial f_0}{\partial V} = - V^p f_1 \tag{12.25}$$

CHAPTER 13. THE MICROSCOPIC MODEL IN A STEADY STATE: NO CONCENTRATION GRADIENTS.

This chapter deals exclusively with the predicted steady state properties of the microscopic kinetic theory model in the absence of any concentration gradients across the membrane. After some preliminary remarks on spatial gradients, I pass to considerations of predicted ionic chord conductance, interionic selectivity, and chord conductance temperature coefficients as functions of applied electric field strength, and ion-scatterer interactions and parameters.

A. Spatial Gradients in a Steady State. First, it is obvious that in a steady state $\dot{f}_0 = \dot{f}_1 = 0$ and $\dot{n} = \dot{j} = 0$. Further, from the continuity equation (6.9), $(\partial j/\partial x) = 0$ and thus $(\partial f_1/\partial x) = 0$. To examine the spatial dependence of f_0 in a steady state, use (12.25) written in the form

$$\frac{\partial f_0}{\partial \bar{X}} - \frac{Z}{V}\frac{d\varphi}{d\bar{X}}\frac{\partial f_0}{\partial V} = -V^p f_1 \tag{13.1}$$

Assume that $f_0(\bar{X},V) = f_0^{\bar{X}}(\bar{X})\, f_0^{V}(V)$ so (13.1) may be written as

$$\frac{df_0^{\bar{X}}}{d\bar{X}} - \frac{d\varphi}{d\bar{X}} f_0^{\bar{X}} \alpha(V, f_0^{V}) = \beta(V, f_0^{V}, f_1) \tag{13.2}$$

wherein

$$\alpha(V, f_0^{V}) = (Z/V)\, d(\ln f_0^{V})/dV \tag{13.3}$$

and

$$\beta(V, f_0^V f_1) = -V^{p-1} f_1/f_0^V \tag{13.4}$$

Furthermore, in a steady state equation 12.24 may be written as

$$-(d\varphi/d\bar{X}) = \gamma(V, f_0^V, f_1) f_0^{\bar{X}} \tag{13.5}$$

where

$$\gamma(V, f_0^V, f_1) = \frac{3V^{p+1}}{2Zf_1}\left(f_0^V + \frac{1}{3V}\frac{df_0^V}{dV}\right) \tag{13.6}$$

Substituting (13.5) into (13.2) immediately gives

$$(df_0^{\bar{X}}/d\bar{X}) + \alpha'(f_0^{\bar{X}})^2 = \beta \tag{13.7}$$

where I have set $\alpha' = \alpha\gamma$. A change of dependent variables in (13.7) by the use of

$$f_0^{\bar{X}} = (1/\alpha')\, d\ln Z/d\bar{X}$$

gives

$$d^2Z/d\bar{X}^2 = \alpha'\beta Z$$

which has the general solution

$$Z(\bar{X}) = A_1 \sin h\sqrt{\alpha'\beta}\,\bar{X} + A_2 \cos h\sqrt{\alpha'\beta}\,\bar{X}$$

Thus, $f_0^{\bar{X}}$ is given by

$$f_0^{\bar{X}} = \sqrt{\frac{\beta}{\alpha'}}\,\frac{A_1 \cos h\sqrt{\alpha'\beta}\,\bar{X} + A_2 \sin h\sqrt{\alpha'\beta}\,\bar{X}}{A_1 \sin h\sqrt{\alpha'\beta}\,\bar{X} + A_2 \cos h\sqrt{\alpha'\beta}\,\bar{X}} \tag{13.8}$$

Assuming that the membrane is of thickness $\bar{\delta}$, i.e. $\bar{X} = \bar{\delta}$ or $x = \delta$, and that the electrolyte solutions on each side of the membrane are identical implies that $f_0^X(\bar{X}=0) = f_0^X(\bar{X}=\bar{\delta})$ and thus from (13.8) $A_1 = A_2$. Therefore when there are no concentration gradients across the membrane, f_0 is independent of $\bar{X}$ and, from (13.5), so is E(constant field).

B. Solutions of the Kinetic Equations for a DC Field.

From the previous section, in a symmetrical situation with an applied electric field across the membrane the dimensionless kinetic equations 12.24 and 12.25 reduce to

$$Z\tilde{E}\,\frac{d(V^2 f_1)}{dV} = \frac{3}{2}\,\frac{d}{dV}\left[\; V^{p+3}\left(f_0 + \frac{1}{3V}\,\frac{df_0}{dV}\right)\right] \tag{13.9}$$

and

$$Z\tilde{E}\,\frac{df_0}{dV} = -V^p f_1 \tag{13.10}$$

respectively. These equations are easily combined to give an equation in f_0 alone;

$$\left[1 + \frac{2(Z\tilde{E})^2}{V^{2p}}\right]\frac{df_0}{dV} + 3Vf_0 = 0 \tag{13.11}$$

which has the solution

$$f_0 = A\,\exp(-W) \tag{13.12}$$

wherein

$$W = \int_0^V \frac{3r\,dr}{1 + 2(Z\tilde{E})^2/r^{2p}} \tag{13.13}$$

It should be noted at this point that f_0 will be Maxwellian if 1) $\tilde{E} = 0$, or 2) $p = 0$, corresponding to an ion-scatterer interaction force with $\alpha=5$.

Thus, from (13.10)

$$f_1 = (Z\tilde{E}A/V^p)(\partial W/\partial V)\,\exp(-W) \tag{13.14}$$

In both (13.12) and (13.14), A is determined from

$$N = 4\pi \int_0^\infty V^2 f_0(V)\,dV \tag{13.15}$$

wherein $N = n/\bar{n}$ as in Chapter 6 and

$$\bar{f}_0 = (f_0/\bar{n})\,v_T{}^3 \tag{13.16}$$

Thus using equation 13.12

$$A = \frac{1}{4\pi \displaystyle\int_0^\infty V^2\,\exp(-W)\,dV} \tag{13.17}$$

In a similar fashion, define $\bar{f}_1 = (f_1/\bar{n})\,v_T{}^3$ so

$$j = (4\pi q/3)\int_0^\infty v^3 f_1\,dv \tag{11.49}$$

may be written as

$$\tilde{I} = (\frac{3-p}{3}) Z\tilde{E} \frac{\int_0^\infty V^{2-p} \exp(-W)\, dV}{\int_0^\infty V^2 \exp(-W)\, dV} \qquad (13.18)$$

from (13.14) and (13.17) with integration by parts.

C. Analytic Behaviour of the Current Density as a Function of Electric Field Strength.

In this section I examine some of the easily determined behaviour of the model membrane current density as a function of electric field strength. To this end define from equation 13.18, as before, two conductance parameters: the chord conductance

$$G_c = \tilde{I}/Z\tilde{E} \qquad (13.19)$$

and the slope conductance

$$G_s = (\partial \tilde{I}/\partial Z\tilde{E}) = G_c + Z\tilde{E}(\partial G_c/\partial Z\tilde{E}) \qquad (13.20)$$

In what follows it will be convenient to set

$$\Delta_1 = \int_0^\infty V^{2-p} \exp(-W)\, dV \qquad (13.21)$$

and

$$\Delta_2 = \int_0^\infty V^2 \exp(-W)\, dV \qquad (13.22)$$

so

$$G_c = \frac{3-p}{3}(\Delta_1/\Delta_2) \qquad (13.23)$$

It is clear by inspection that G_c is an even function of $Z\tilde{E}$. Further for small $\tilde{E}$, $W \rightarrow (3V^2/2)$ and by use of the integral definition of the gamma function

$$\int_0^\infty x^{\nu-1} e^{-\mu x} dx = \mu^{-\nu}\Gamma(\nu), \quad \mathrm{Re}(\mu,\nu) > 0 \qquad (13.24)$$

I may write the chord conductance for vanishingly small $\tilde{E}$ as

$$G_c(\tilde{E}=0) = (\frac{3-p}{3})(\frac{3}{2})^{p/2} \frac{\Gamma(\frac{3-p}{2})}{\Gamma(\frac{3}{2})} \qquad (13.25)$$

It is to be noted that for p=0 (corresponding to ion-induced dipole interactions) and all $\tilde{E}$, $\Delta_1=\Delta_2$ and therefore $G_c = 1$ to predict totally ohmic system behaviour.

To examine the behaviour of $\partial G_c/\partial(Z\tilde{E})$ use (13.23) to give

$$\frac{\partial G_c}{\partial(Z\tilde{E})} = \left(\frac{3-p}{3}\right)\frac{\Delta_1}{\Delta_2}\left[\frac{1}{\Delta_1}\frac{\partial\Delta_1}{\partial(Z\tilde{E})} - \frac{1}{\Delta_2}\frac{\partial\Delta_2}{\partial(Z\tilde{E})}\right] \tag{13.26}$$

wherein

$$\frac{\partial\Delta_1}{\partial(Z\tilde{E})} = -\int_0^\infty V^{2-p}(\partial W/\partial Z\tilde{E})\exp(-W)\,dV \tag{13.27}$$

$$\frac{\partial\Delta_2}{\partial(Z\tilde{E})} = -\int_0^\infty V^{2}(\partial W/\partial Z\tilde{E})\exp(-W)\,dV \tag{13.28}$$

and

$$\frac{\partial W}{\partial(Z\tilde{E})} = -4Z\tilde{E}\int_0^V \frac{3r\,dr}{r^{2p}\left[1 + 2(Z\tilde{E})^2/r^{2p}\right]^2} \tag{13.29}$$

For small $\tilde{E}$, $W \to 3V^2/2$ so Δ_1 and Δ_2 are independent of $\tilde{E}$ and $\partial G_c/\partial(Z\tilde{E}) = 0$ as expected. However, for large $\tilde{E}$

$$W \to V^{2(p+1)}/4(p+1)(Z\tilde{E})^2 = \mu_1 V^{2(p+1)} \tag{13.30}$$

where

$$\mu_1 = 1/4(p+1)(Z\tilde{E})^2 \tag{13.31}$$

and Δ_1 and Δ_2 become

$$\Delta_1 = \Gamma(\nu_1)/2(p+1)\mu_1^{\nu_1} \qquad p\neq -1 \tag{13.32}$$

and

$$\Delta_2 = \Gamma(\nu_2)/2(p+1)\mu_1^{\nu_2} \qquad p\neq -1 \tag{13.33}$$

respectively, with

$$\nu_1 = (3-p)/2(p+1), \qquad \nu_2 = 3/2(p+1) \tag{13.34}$$

Also,

$$\frac{\partial\Delta_1}{\partial(Z\tilde{E})} = -\mu_2\Gamma(\nu_1+1)/2(p+1)\mu_1^{(\nu_1+1)} \tag{13.35}$$

and

$$\frac{\partial \Delta_2}{\partial (Z\tilde{E})} = -\mu_2 \Gamma(\nu_2+1)/2(p+1)\mu_1^{(\nu_2+1)} \qquad (13.36)$$

where

$$\mu_2 = -2\mu_1/Z\tilde{E} \qquad (13.37)$$

Combining the results of equations 13.31-13.37 with 13.23 gives an approximate high field expression for G_c:

$$G_c = (\frac{3-p}{3})\left[4(p+1)(Z\tilde{E})^2\right]^{-p/2(p+1)} \Gamma(\nu_1)/\Gamma(\nu_2) \qquad (13.38)$$

Thus for $p > 0$, G_c is a decreasing function of $Z\tilde{E}$, for $p=0$ it is constant and identically equal to 1 as shown above, while for $-1 < p < 0$ G_c is an increasing function of $Z\tilde{E}$.

These characteristics of G_c variation with $Z\tilde{E}$ are all intuitively as expected. The model contains no inherent asymmetry for the situation of no concentration gradient, and thus no asymmetry in G_c with respect to the sign of $Z\tilde{E}$ is expected. On a simple basis, G_c should be inversely proportional to the collision frequency - the more collisions an ion makes in going through a region the higher the resistance of that region, and thus the lower its conductance. The effect of an external electric field is to increase the velocity of an ion. For $p > 0$ (ν an increasing function of V) this in turn increases the collision frequency and thus decreases the conductance. Conversely, for $p < 0$ (ν a decreasing function of V) an increase in the field strength decreases the collision frequency and increases the conductance. For a velocity independent collision frequency ($p=0$) the electric field strength has no effect on ν and also no effect on the conductance.

Returning to (13.26), it is easy to show that in the high field approximation

$$\frac{\partial G_c}{\partial (Z\tilde{E})} = -(\frac{3-p}{3})\frac{\Delta_1}{\Delta_2} \cdot \frac{p}{(Z\tilde{E})(p+1)}$$

$$= -(\frac{3-p}{3})\frac{p}{(Z\tilde{E})(p+1)}\left[4(p+1)(Z\tilde{E})^2\right]^{-p/2(p+1)} \Gamma(\nu_1)/\Gamma(\nu_2) \qquad (13.39)$$

to predict that the high field slope conductance should be given by

$$G_s = \frac{3-p}{3(p+1)}\left[4(p+1)(Z\tilde{E})^2\right]^{-p/2(p+1)} \Gamma(\nu_1)/\Gamma(\nu_2)$$
$$= \frac{2}{3}\left[4(p+1)(Z\tilde{E})^2\right]^{-p/2(p+1)} \Gamma(\nu_1+1)/\Gamma(\nu_2) \qquad (13.40)$$

Thus it is easy to see that the slope conductance at high values of the applied electric field is always non-negative, and further if (13.38) is written in the form

$$G_c = \frac{2(p+1)}{3}\left[4(p+1)(Z\tilde{E})^2\right]^{-p/2(p+1)} \Gamma(\nu_1+1)/\Gamma(\nu_2) \qquad (13.41)$$

then $G_c = (p+1)G_s$ for $|Z\tilde{E}| >> 0$.

D. Computed Behaviour of $\tilde{I}(\tilde{E})$ and $G_c(\tilde{E})$.

As seen in the previous section the complexity of equations 13.18-13.19 for $\tilde{I}(\tilde{E})$ and $G_c(\tilde{E})$ makes a complete analytical study of the characteristics of ion movement through this simple membrane model difficult; a numerical study is therefore appropriate. From considerations in Chapter 11, $\nu \propto v^p$, and so the results of these computations are presented in two sections, the first for $p < 0$ and the second for $p > 0$. The p=0 case, with ohmic conductance characteristics, provides a natural dividing line.

Computations of the conductance and current-electric field curves were carried out for a range of p values between -0.5 and -3.0. However only two values correspond to actual ion scattering interactions, p = -3 and p = -1. In terms of ideal interactions these would correspond to ion-fixed charge (α=2) and ion-permanent dipole (α=3) interactions respectively. The computed G_c versus $\tilde{E}$ curves are shown in Figure 13.1 and the $\tilde{I}$ versus $\tilde{E}$ curves are in Figure 13.2.

As expected from the general considerations of the previous section, and illustrated in Figure 13.1, G_c is an increasing function of $\tilde{E}$ for those ion-scattering interactions in which the collision frequency decreases with increasing ionic velocity. The non-linear field dependent behaviour of G_c for these types of interactions becomes much less pronounced at high field strengths, the conductance varies slowly with $\tilde{E}$, and the $\tilde{I}$ versus $\tilde{E}$ curves becomes nearly linear. For p = -3 the increase in conductance with field strength is rapid, leveling off for high fields at a value (125) about one

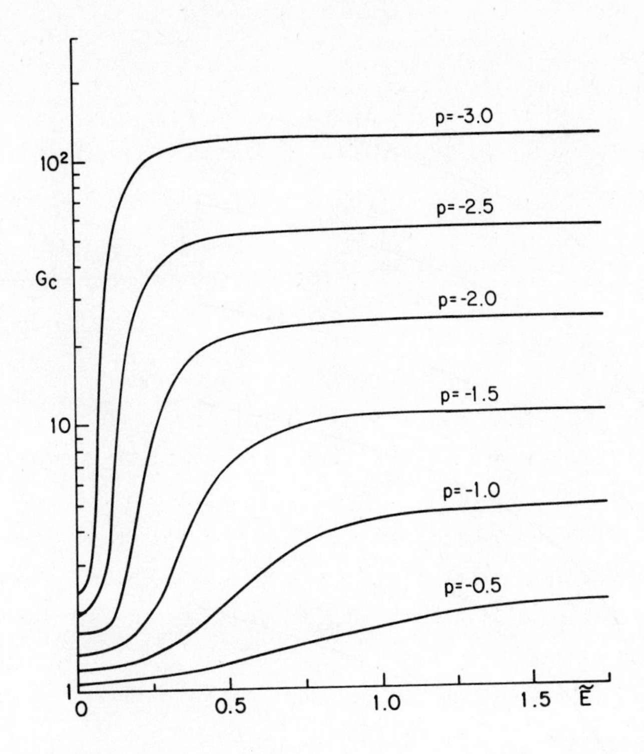

Figure 13.1 Dimensionless conductance, G_c, as a function of dimensionless electric field strength, $\tilde{E}$, for a number of ion-scatterer interactions characterized by $p < 0$. The curves p=-1 and p=-3 correspond to ion-fixed dipole and ion-fixed charge interactions respectively.

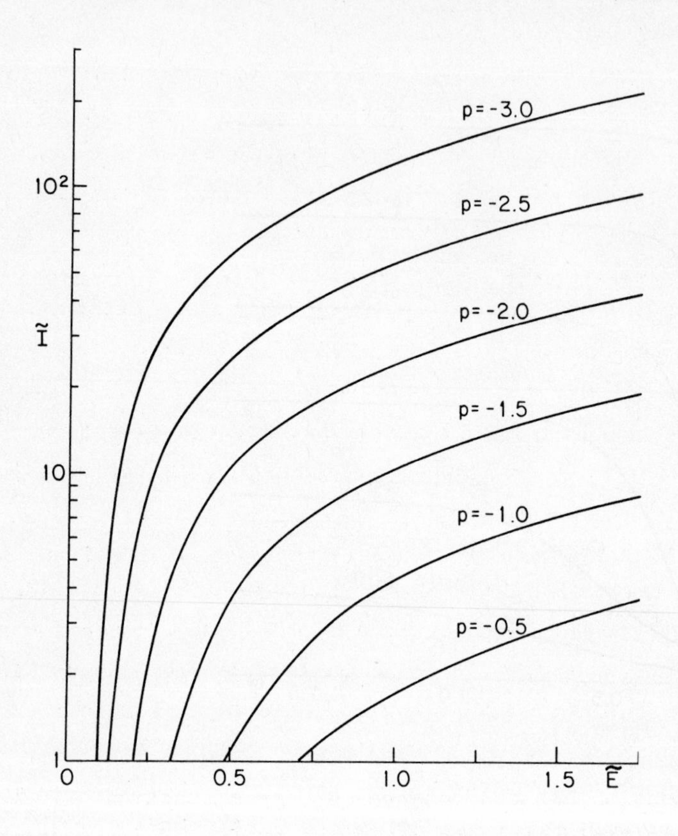

Figure 13.2 Dimensionless current, $\tilde{I}$, as a function of the dimensionless electric field strength, $\tilde{E}$, for six ion-scatterer interactions with $p < 0$. The computed curves are presented semi-logarithmically for clarity.

and one-half orders of magnitude greater than its value at zero field strength (about 2.44).

When p=0 (corresponding, ideally, to ion-induced dipole interactions), G_c=1, and the dimensionless $\tilde{I}$ versus $\tilde{E}$ curve is a straight line with unity slope. This is intuitively reasonable. When the collision frequency is independent of ionic velocity, and thus also of electric field strength, the conductance is constant.

Two cases were examined numerically for $p > 0$, p=1/3 and p=1. A p of 1/3 corresponds to a dominant induced dipole-induced dipole (α=7, London dispersion force) interaction, which seems an unlikely mechanism to be operating during ion movement through a membrane. A p=1 results if a classical hard sphere-hard sphere collision process is assumed, i.e., collision of an ion with a non-polarizable, non-charged scatterer.

In Figure 13.3 the dimensionless conductance G_c for p=1/3 and p=1 is plotted versus $\tilde{E}$. In both cases, G_c is a decreasing function of $\tilde{E}$ and for p=1 the conductance is a more rapidly decreasing function of $\tilde{E}$ than it is for p=1/3. The consequences of these differences for the dimensionless current versus electric field curves are illustrated in Figure 13.4.

If a selectivity coefficient, S_i, for the i th ionic species is defined as the ratio of actual current carried by the i th species relative to the actual current carried by a reference species, j, at constant field strength, then

$$S_i = \tilde{I}_i/\tilde{I}_j \tag{13.42}$$

From the definitions of Chapter 6, and equation 13.18, (13.42) may be written out explicitly as

$$S_i = \frac{Z_i}{Z_j}\left(\frac{m_j+m_s}{m_i+m_s}\right)^{\frac{1}{2}} \eta_{ij} \frac{G_c(\eta_{ij}\tilde{E}_j)}{G_c(\tilde{E}_j)} \tag{13.43}$$

wherein

$$\eta_{ij} = \frac{Z_i}{Z_j}\left(\frac{K_{js}}{K_{is}}\right)^{\frac{1-p}{2}} \left(\frac{m_j}{m_i}\right)^{\frac{1}{2}} \left(\frac{m_i+m_s}{m_j+m_s}\right)^{\frac{p}{2}} \tag{13.44}$$

Table 13.1 shows the ratio of the force constants, K_{js}/K_{is}, for the interactions considered, and all other symbols have been previously defined.

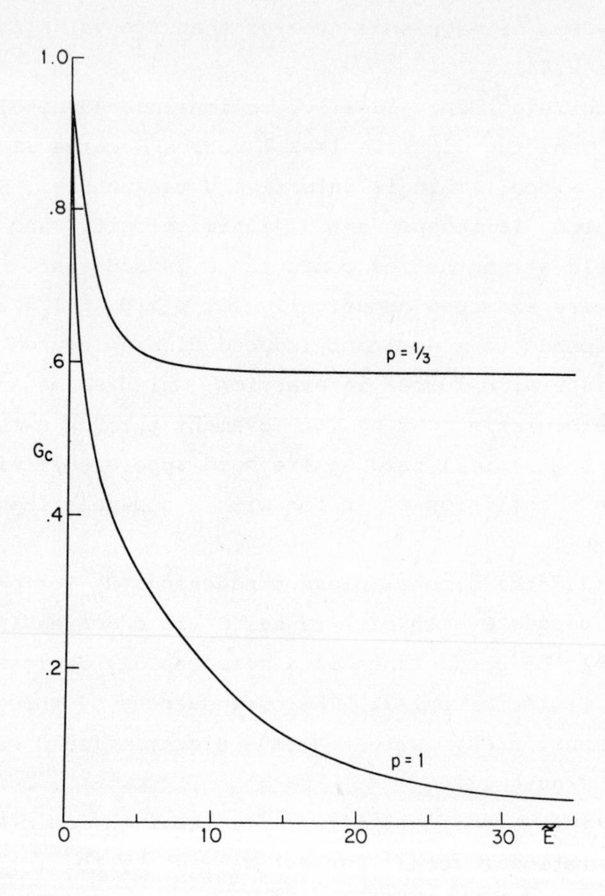

Figure 13.3 Dimensionless conductance as a function of the dimensionless electric field strength for classical ion-induced dipole and ion-neutral scattering interactions, characterized by p = 1/3 and p = 1 respectively.

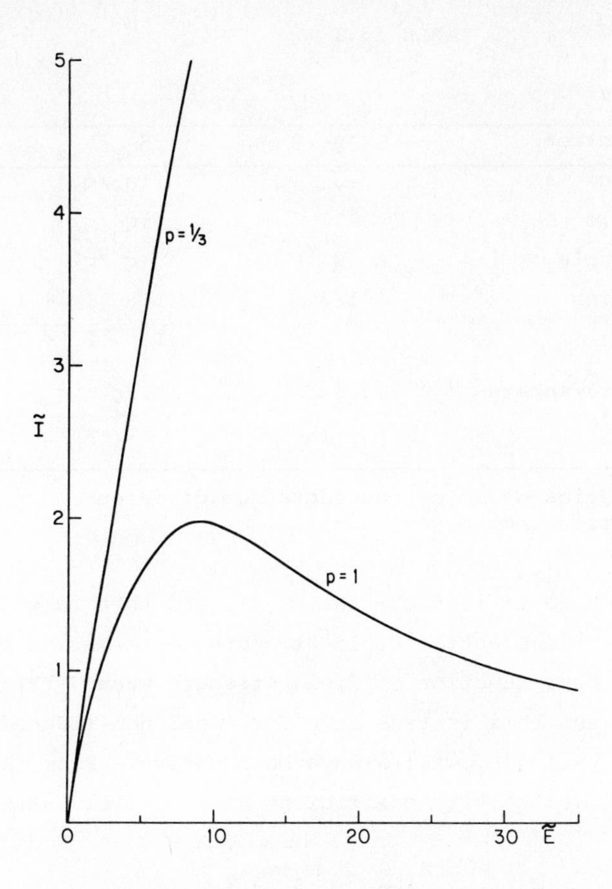

Figure 13.4 The dimensionless current versus dimensionless electric field strength relations for the ion-scatterer interactions of Figure 13.3.

TABLE 13.1

Type of interaction	p	K_{js}/K_{is}
Ion-fixed charge	-3	(q_j/q_i)
Ion-fixed dipole	-1	(q_j/q_i)
Ion-induced dipole	0	(q_j/q_i)
London dispersion	1/3	$\frac{\alpha_j I_j}{\alpha_i I_i} - \frac{I_i + I_s}{I_j + I_s}$
Hard sphere-hard sphere	1	$\frac{r_j + r_s}{r_i + r_s}$

Table 13.1. Ratios of force constants for different idealized interactions.

From these formulae it is clear that $S_j=1$, and that in general S_i is a function of the electric field strength, or membrane potential. S_i will not be a function of field strength when $G_c(\eta_{ij}\tilde{E}_j)= G_c(\tilde{E}_j)$ for all $\tilde{E}_j$, and this is true only for ideal non-induced dipole interactions (p=0). S_i will always be positive. From (13.43) it follows that the selectivity coefficient at zero field strength is

$$S_j(E_j=0) = \frac{Z_i}{Z_j}\eta_{ij}\left(\frac{m_j+m_s}{m_i+m_s}\right)^{\frac{1}{2}}$$

Also,

$$\lim_{E_j \to \infty} G_c(\eta_{ij}\tilde{E}_j) = \lim_{E_j \to \infty} G_c(\tilde{E}_j)$$

so

$$\lim_{E_j \to \infty} S_i(\tilde{E}_j) = S_i(\tilde{E}_j=0)$$

Thus, no matter what the variation of selectivity with field strength the selectivity for very high fields will always approach the value it had at zero field strength.

In the numerical evaluation of S_i sodium is taken as the standard species, $S_{Na}=1$. For every interaction, the selectivity coefficient has been calculated for Li^+, NH_4^+, K^+, Rb^+, and Cs^+ since these are the most commonly studied ions in selectivity studies on excitable membranes. The parameters needed for the computation (e.g., ionic mass, radius, and polarizability) are given in Table 13.2. These values, it must be stressed, are for unhydrated ions.

TABLE 13.2

Ion	Ionic constants			
	Molecular weight	Polarizability (α) x 10^{24}	Second Ionization potential(I)	Crystal radius
		cm^2	ev	A
Li^+	6.940	0.075	75.28	0.68
Na^+	22.997	0.21	47.07	0.98
NH_4^+	18.040	1.65	31.70	1.45
K^+	39.100	0.87	31.70	1.33
Rb^+	85.480	1.81	27.30	1.48
Cs^+	132.91	2.79	23.40	1.67
Scatter(carboxyl oxygen)		0.84	2.12	1.45

References: Conway (1952); Handbook of Chemistry and Physics (1957); Ketelaar (1953); Latimer (1952); Moelwyn-Hughes (1949); Mulliken (1933).

Table 13.2. Molecular properties of unhydrated cations needed for various computations.

Much has been written speculating about the hydration state of ions crossing membranes, but as of this time there is little evidence either supporting or rejecting the hypothesis that ions go through membranes in hydrated form. Ling (1962) has summarized a number of studies which attempt to determine the state of hydration of several monovalent cations in free solution. The most consistent feature of these estimates is their variability. For example, estimates of the hydrated radius of sodium range from 1.74A to 7.90A and the hydration number from 1 to 180. Clearly, just about anything could be explained by using published data on the hydration state of an ion. For this reason only unhydrated parameters have been used.

Figure 13.5 is a graph of the selectivity S_i versus field strength $\tilde{E}_{Na}$ for the six monovalent cations listed above for ion-

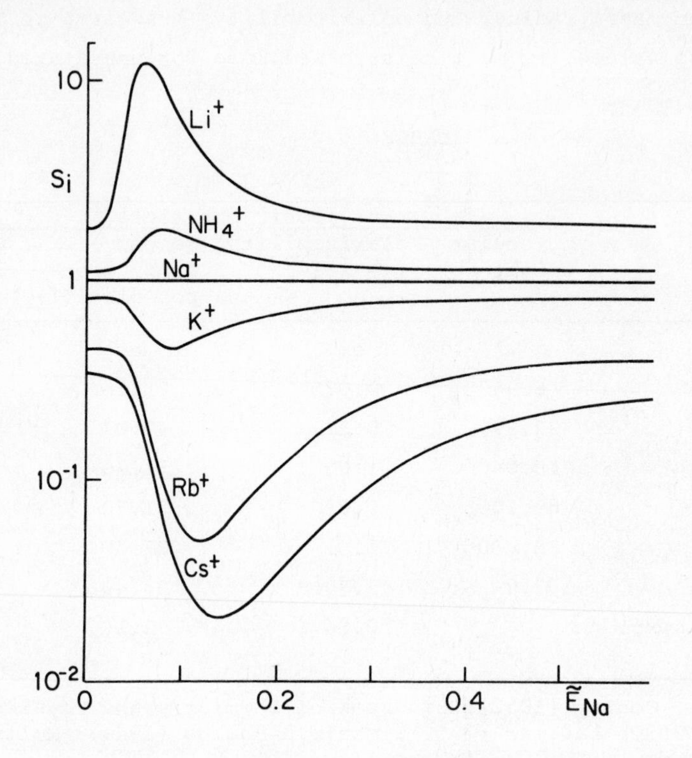

Figure 13.5 Selectivity, S_i, versus dimensionless electric field strength, $\tilde{E}_{Na}$, for i = Li^+, NH_4^+, Na^+, K^+, Rb^+ and Cs^+ and ion-fixed charge interactions (p = -3). Effective scatterer molecular weight is 1000. Note that ions with $S_i > 1 (< 1)$ have maxima (minima) in their S_i versus field strength curves. For any interaction such that $p < 0$, qualitatively identical results are found.

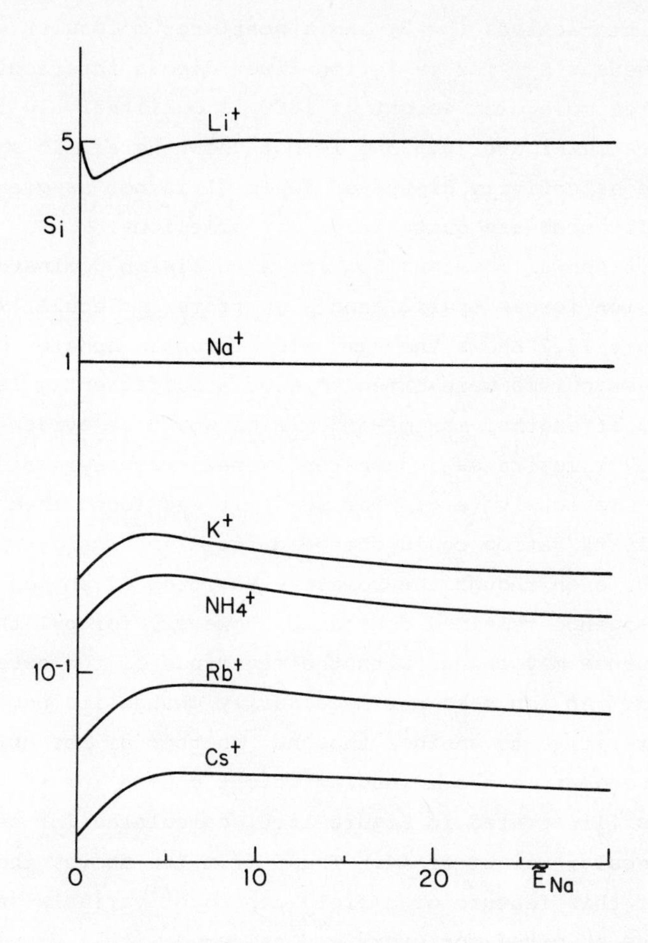

Figure 13.6 Selectivity as a function of dimensionless electric field strength for six different cations experiencing London dispersion force ($p = 1/3$) interactions as they traverse the membrane. Scatterer molecular weight is 1000. Note that ions whose $S_i < 1$ (>1) exhibit minima (maxima) in their S_i versus $\tilde{E}_{Na}$ curves.

fixed charge interactions (p=-3) and a scatterer molecular weight of 1000. S_i versus $\widetilde{E}_{Na}$ for p=-1 (ion-fixed dipole interaction) with a scatterer molecular weight of 1000 is qualitatively identical with the curves of Figure 13.5 and is not shown here. It should be noted that the selectivity displayed for p=-1 is not as great as for p=-3; the differences are quite large for some ions.

Figure 13.6 shows S_i versus $\widetilde{E}_{Na}$ for a collision dominated by London dispersion forces (p=1/3) and a scatterer molecular weight of 1000. Figure 13.7 shows the same plot for hard sphere collisions (p=1). If these curves were extended over a sufficiently large range of field strengths, maxima and minima would be evident.

Figure 13.7 illustrates an apparently new characteristic of the selectivity. Previously (e.g., for p=-3) it was found that the relative selectivity ratios could change quite dramatically with the field strength, even though the relative position of an ion in the selectivity sequence remained constant. However, for p=1 the selectivity sequence may change with the magnitude of the external electric field. An ion need not necessarily change its position in the sequence relative to another though. Whether or not this happens is dependent on the molecular weight chosen for the scattering center as illustrated in Figure 13.8, calculated for p=1 and a scatterer molecular weight of 44. Thus, from the above, the question arises if this feature of a field dependent variable selectivity sequence can be observed for every p considered with a proper choice of parameters. This is not the case. Consider first of all the situation when $p < 0$. Suppose that two monovalent ions i and k might exchange relative positions in a selectivity sequence. This is equivalent to saying that for some $\widetilde{E}_{Na}$,

$$S_i(\eta_{is}\widetilde{E}_s) = S_k(\eta_{ks}\widetilde{E}_s)$$

or equivalently

$$\frac{\eta_{is}}{\eta_{ks}}\left[\frac{m_k+m_j}{m_i+m_j}\right]^{\frac{1}{2}} = \frac{G_c(\eta_{ks}\widetilde{E}_s)}{G_c(\eta_{is}\widetilde{E}_s)} \tag{13.45}$$

Distinguish two different cases:

Case 1. $\underline{G_c(\eta_{ks}\widetilde{E}_s) < G_c(\eta_{is}\widetilde{E}_s)}$. From the properties of $G_c(\widetilde{E}_s)$ for $p < 0$, this is equivalent to saying

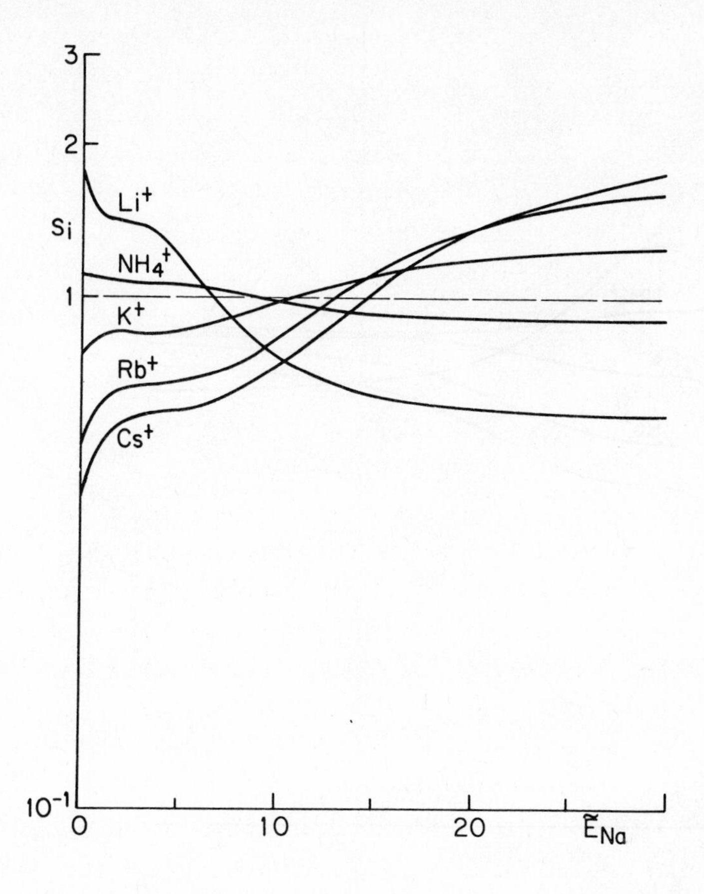

Figure 13.7 Ionic selectivity versus electric field strength for six cations, ion-neutral scatterer (p = 1) interactions, and a scatterer molecular weight of 1000. Note that there exist 11 different selectivity sequences, depending on the value of the field strength. Extension of the curves for larger values of field strength would demonstrate the characteristic maxima and minima.

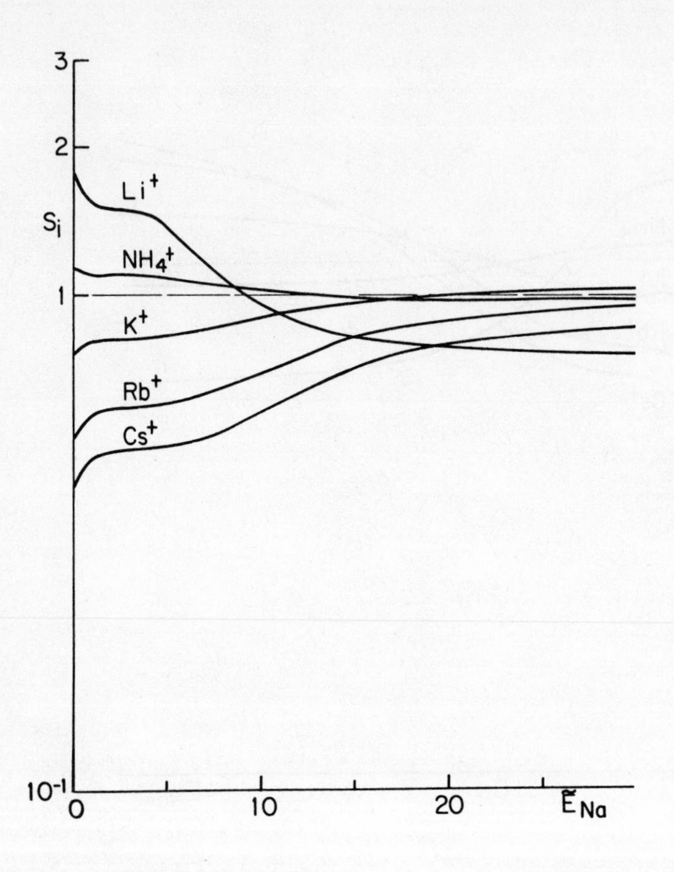

Figure 13.8 Selectivity as a function of electric field strength as in Figure 13.7, but with a scatterer molecular weight of 44 to illustrate the effect of varying this parameter. The 11 sequences of Figure 13.7 have been reduced to six for this range of field strengths.

$$\eta_{ks} < \eta_{is} \tag{13.46}$$

But, from (13.45) we must also have

$$\frac{\eta_{is}}{\eta_{ks}}\left[\frac{m_k+m_j}{m_i+m_j}\right]^{\frac{1}{2}} < 1$$

or with (13.46)

$$\left[\frac{m_k+m_j}{m_i+m_j}\right]^{\frac{1}{2}} < 1$$

or

$$m_k < m_i \tag{13.47}$$

Returning to (13.46) and inserting the full expressions for η_{is} and η_{ks}, we find that this is equivalent to

$$\left(\frac{K_k}{K_i}\right)^{\frac{1-p}{2}}\left(\frac{m_k}{m_i}\right)^{\frac{1}{2}}\left[\frac{m_i+m_j}{m_k+m_j}\right]^{\frac{p}{2}} > 1 \tag{13.48}$$

Because $p < 0$ and $m_k < m_i$ by (13.47), (13.48) reduces to

$$K_i > K_k$$

as a necessary but not sufficient condition for the altering of a sequence. But, for the forces that have been dealt with in which $p < 0$, $K_i = K_k$ for all ions. Thus we have a contradiction and for $p < 0$ a selectivity sequence alteration is impossible.

Case 2. <u>$G_c(\eta_{ks}\tilde{E}_s) > G_c(\eta_{is}E_s)$</u>. This may be followed through in a manner identical with that of Case 1 to again reach a contradiction.

Thus for $p < 0$ (or more specifically, $p = -1$ or $p = -3$), it is impossible for an alteration in a selectivity sequence to occur with any given set of ionic or scatterer parameters.

The same simple types of arguments may be carried through for positive p, and it is trivially obvious that one may obtain selectivity sequence alterations if ionic parameters are approximately picked for $p=1/3$ and $p=1$.

Experimentally, consideration is not usually given to variation in selectivity with membrane potential. Experimental determinations of selectivity in membranes are often based on the effects of ion substitution on the equilibrium potential of a particular pathway. From the observed shift in the equilibrium potential a permeability ratio is obtained from the Hodgkin-Katz equation. This is a different measure of the selectivity of a system than is S_i.

Appropriate experimental data might be obtained from measurements of current as a function of membrane potential taken, for example, on a squid giant axon perfused with high equal potassium concentrations both inside and out, and then the same experiment repeated with a number of different cations replacing potassium. Unfortunately, such explicit data do not exist in the published literature. However, there is some data indicating that S_i may be a function of membrane potential. Tables 13.3 and 13.4 present data taken from Chandler and Meves (1965) and Adelman and Senft (1966) respectively. Both measured the effect of replacing internal potassium by rubidium or cesium on the delayed outward currents through the potassium channel. There does seem to be a variation of both S_{Rb} and S_{Cs} with membrane potential; both selectivity coefficients are less than one for zero membrane potential, and show minima as the membrane potential is increased.

TABLE 13.3

V_m(mV)	S_{Rb}
-25	.50
0	.26
25	.25
50	.30
75	.34

Table 13.3. Selectivity coefficient versus membrane potential for rubidium ions going through the potassium channel. After Chandler and Meves (1965).

The mechanism of intra-ionic selectivity in this model of membrane ion transport has one basic feature: selectivity is determined by the same process that gives rise to non-linear electric field dependent conductances, i.e., ion-membrane molecule interactions. Two aspects of this interaction important for selectivity

TABLE 13.4

V_m(mV)	S_{Cs}
-40	.60
-20	.42
0	.28
20	.21
40	.12
60	.097
80	.091
92	.110

Table 13.4. Selectivity coefficient versus membrane potential for cesium ions traversing the potassium channel. After Adelman and Senft (1966).

are (1) the dependence of the (classical) ion-scatterer interaction on intra-particle separation and (2) the ionic and molecular masses and parameters entering the (classical) force constant, K_{is}. The coupling of a selectivity mechanism to a non-linear conductance seems to be novel, though all previous selectivity theories have been based on variations of ionic and/or molecular parameters. However, the highly non-linear way in which these parameters may determine selectivity does not seem to have been considered.

The three qualitative generalizations that may be made about the selectivity of the model system are: (1) The selectivity coefficient is a function of the electric field strength except for p=0; (2) As a consequence of the dependence of chord conductance on field strength for $p < 0$, ions whose $S_i > 1 (< 1)$ at zero field strength will show maxima (minima) in their S_i versus $\tilde{E}_j$ curves. Conversely for $p > 0$ ions with $S_i > 1(< 1)$ at zero field strength will have minima (maxima) in their S_i versus E_j curves; (3) For $p > 0$, the sequence of selectivities may be altered by changes in field strength.

Temperature, like ionic and scatterer parameters, was lumped in the dimensionless current, conductance, and electric field calculations. Hence a temperature variation coefficient, R_T, for the theoretical currents may be defined:

$$R_T = \frac{\tilde{I}(T)}{\tilde{I}(T_s)}$$

where T_s is a standard temperature in degrees Kelvin. Proceeding as in the derivation of (13.43) the formula

$$R_T = \left(\frac{T}{T_s}\right)^{\frac{1}{2}} \delta \frac{G_c(\delta \tilde{E}_s)}{G_c(\tilde{E}_s)} \tag{13.49}$$

is obtained, wherein

$$\delta(T) = \left(\frac{T_s}{T}\right)^{\frac{p+1}{2}}$$

Thus from 13.49 the theoretical Q_{10} of the current, which is $R_T(T = T_s + 10)$, is given by

$$Q_{10} = \left(\frac{T_s}{T_s+10}\right)^{\frac{p}{2}} \frac{G_c(\delta \tilde{E}_s)}{G_c(\tilde{E}_s)} \tag{13.50}$$

Generally the Q_{10} of the model currents will be a function of electric field strength, but there are two important special cases for which this does not hold. When p=0 (ion-induced dipole scattering) the Q_{10} has the constant value of one (from (13.50) and the properties of G_c for p=0). The other special case occurs for ion-permanent dipole scattering (p=-1). In this instance $\delta(T) = 1$ and

$$Q_{10} = \left(\frac{T_s+10}{T_s}\right)^{\frac{1}{2}}$$

For a standard temperature, T_s, of 273°K the Q_{10} is 1.01815 while for a T_s = 300°K it is 1.01653.

The variation of the Q_{10} for the currents with electric field strength is illustrated in Figure 13.9 for ion-fixed charge scattering interactions (p=-3). The Q_{10} rises rapidly from a zero field value of about 1.055 to a peak around 1.189, and then falls somewhat more slowly to nearly attain its initial value at a $\tilde{E}_s$ = 0.9. For induced dipole-induced dipole interactions (p=1/3), the Q_{10} versus $\tilde{E}_s$ curve illustrated in Figure 13.10 is qualitatively similar to that for p=-3. There are, however, two differences. (1) The temperature coefficient is less than one for all $\tilde{E}_s$, and (2) The maximum change in the Q_{10} is considerably less (about 5 parts in 1000) than that observed for p=-3 (135 parts in 1000).

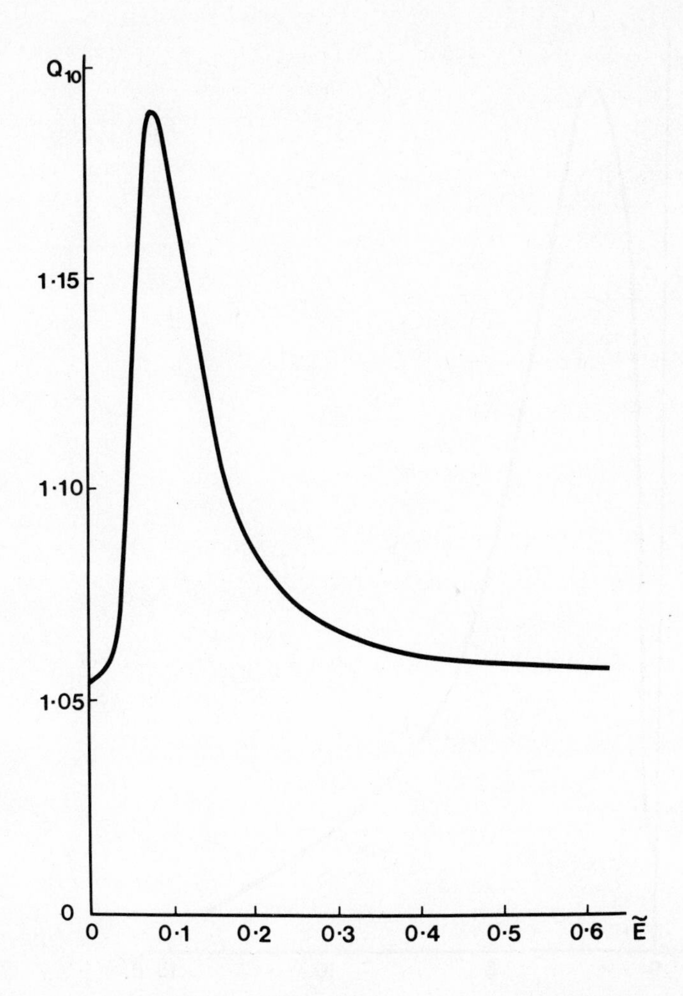

Figure 13.9 Temperature coefficient versus dimensionless electric field strength for ion-fixed charge interactions. Current Q_{10} versus $\tilde{E}$ for ions moving through a negative fixed charge scattering medium (p = -3), initial temperature is 273°K.

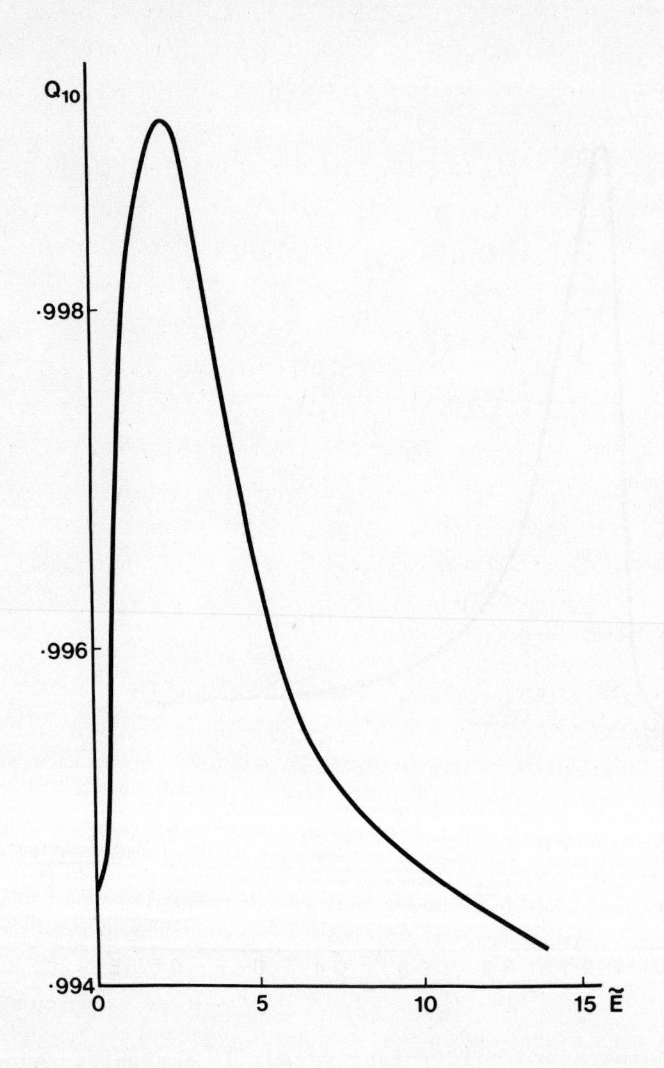

Figure 13.10 Temperature coefficient versus electric field strength for induced dipole-induced dipole interactions. The initial temperature was 273°K.

An increase in temperature is roughly equivalent to an increase in electric field strength since both tend to increase the energy, and hence the velocity,of an ion. Thus for $p < 0$, where a temperature induced ionic velocity increase would decrease the average collision frequency, a $Q_{10} > 1$ is found. Correspondingly for p=0, and thus an ohmic conductance, the temperature coefficient is one. For interactions giving an increase in collision frequency with an increase in particle velocity ($p > 0$) a $Q_{10} < 1$ is found. This is as was found for p=1/3. For p=1, this simplistic concept holds only for small values of $\tilde{E}_s$, as shown in Figure 13.11 but fails for larger values of $\tilde{E}_s$. In the previously considered cases the factor $(T_s/(T_s+10))^{p/2}$ reflects the general qualitative nature of the Q_{10} ($\gtrless 1$) while the factor $G_c(\delta \tilde{E}_s)/G_c(\tilde{E}_s)$ is a quantitative modulation of this. In the case p=1, however, the rate of chord conductance decrease with increasing temperature is sufficient to reverse the effects of the first factor because $G_c(\delta \tilde{E}_s) > G_c(\tilde{E}_s)$ for all $|\tilde{E}_s| > 0$.

E. Estimation of Some Membrane Related Quantities.

It is instructive to examine actual current densities, electric field strengths, and conductances to see how theoretical predictions match experimental fact. Thus, the following hypothetical membrane is postulated to permit comparison of the calculated $\tilde{I}$ versus $\tilde{E}$ and G_c versus $\tilde{E}$ relationship with experimental ones.

The model membrane is a planar sheet of non-conducting lipid and protein 100 A in thickness, with ion permeable regions whose molecular characteristics are one of those employed herein for calculation; e.g., fixed charges, polarizable particles, etc. Assume that the scatterers in these conducting regions are components of the surrounding lipid-protein matrix; that these regions may be approximated by cylinders 5 A in diameter extending through the membrane, and that there are 10 of these regions per square micron of membrane surface. The estimate of 100 A for the membrane thickness is consistent with the measured thickness (Chapter 1) and the density of pores is based on work with TTX (Chapter 5).

Another quantity that must be estimated is the number density of ions within the membrane. If the hypothetical membrane has a maxi-

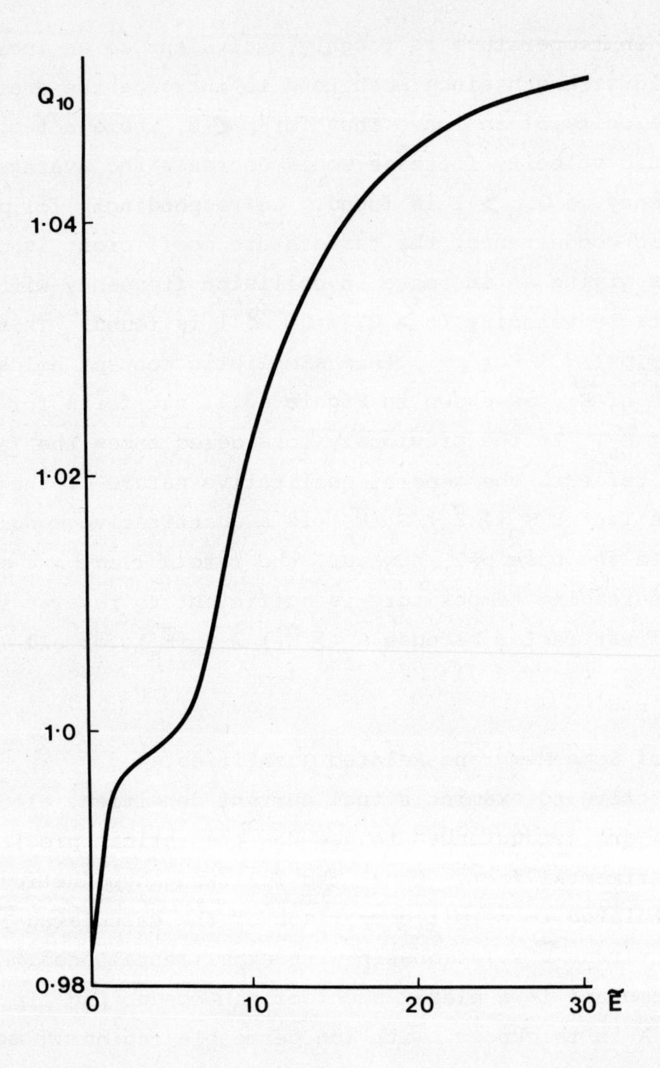

Figure 13.11 Temperature coefficient versus dimensionless electric field strength for ion-neutral particle (p = 1) interactions with an initial temperature of 273°K. Note that $Q_{10} < 1$ for small $\tilde{E}$, and then becomes greater than one as $\tilde{E}$ increases. If the curve was extended to larger $\tilde{E}$, a maximum would be observed.

mum current density of 1 mA/cm^2 flowing through it, carried by monovalent cations, then it is easy to show that this corresponds to 6.25×10^6 ions/sec through each pore. For a membrane bathed in 0.1 molar electrolyte, elementary kinetic theory indicates that the cross sectional area of one pore would suffer 6.4×10^9 collisions every second. Thus only about one ion in 10^3 goes through the pore, and it will therefore be assumed that the number density of ions interacting with the scatterers is 10^{-3} that of the external solution.

Another necessary quantity is the number density of scattering centers. This is impossible to estimate, so it was arbitrarily taken as $6 \times 10^{20}/cm^3$, or about one per pore. As estimates of the effective mass of the scattering center two values were taken, using molecular weights of 44 and 1000. The value of 44 was taken because it is the molecular weight of a carboxyl group. Using the parameters of the sodium ion the energy loss factor, ξ_{Na}, has the value of about 0.5 for this effective scattering mass which is large for the approximations made earlier. Thus a second effective molecular weight of 1000 (making $\xi_{Na} = 0.044$) was also used. This effective mass is reasonable if the molecule to which the scattering center belongs is quite rigid. Various interactions require estimates about the charge of the scattering center, or its polarizability, etc. For ion-fixed charge collisions it was assumed that the scatterer carries a net charge of -1, while for ion-fixed permanent dipole interactions a dipole moment of 5D was assumed. For ion-induced dipole interactions, induced dipole-induced dipole interactions, and ion-neutral scatterer interactions the necessary properties for the sodium ion and carboxyl oxygen (scatterer) from Table 13.2 were used.

In Table 13.5 are presented, for various interactions, the membrane potential, V_m, (assuming a constant electric field) corresponding to a given $\tilde{E}_{Na}$; the current density per pore, j_p, and current density per square centimeter of membrane, j_m, corresponding to a given $\tilde{I}_{Na}$, and the conductance per pore, G_p, and per square centimeter of membrane, G_m. The first entry in each bin of the table was calculated with an effective scatterer molecular weight of 44, the second with a molecular weight of 1000. The values given may be

TABLE 13.5

p	$V_m(\tilde{E}=1)$ mV	$j_p(\tilde{I}=1)$ A/pore	$G_p(G_c=1)$ mho/pore	$j_m(\tilde{I}=1)$ mA/cm^2	$G_m(G_c=1)$ mmho/cm^2
-3	1.5×10^{7} (MW=44)	8.6×10^{-12}	5.7×10^{-19}	8.6	5.7×10^{-7}
	1.7×10^{6} (MW=1000)	2.2×10^{-12}	1.3×10^{-18}	2.2	1.3×10^{-6}
-1	6.8×10^{4}	as above	1.3×10^{-16}	as above	1.3×10^{-4}
	1.2×10^{4}	as above	1.8×10^{-16}	as above	1.8×10^{-4}
0	5.5×10^{2}	as above	1.6×10^{-14}	as above	1.6×10^{-2}
	1.2×10^{2}	as above	1.8×10^{-14}	as above	1.8×10^{-2}
1/3	99	as above	8.7×10^{-14}	as above	8.7×10^{-2}
	23	as above	9.6×10^{-14}	as above	9.6×10^{-2}
1	44	as above	1.96×10^{-13}	as above	.196
	11	as above	2.0×10^{-13}	as above	.20

Table 13.5. Model specific conductances, currents, and membrane potentials for different intermolecular interactions and sparsely distributed pores. For each interaction the membrane potential (V_m) current density per pore (j_p) and per square centimeter of membrane (j_m); and, conductance per pore (G_p) and square centimeter of membrane G_m) corresponding to unit values of the dimensionless variables $\tilde{E}$, $\tilde{I}$, and G_c are given. For example, with ion-neutral particle scattering and a scatterer MW of 1000, an $\tilde{E}$=1 corresponds to a V_m=11 mV, and $\tilde{I}$=1 corresponds to a I_m=2.2 mA/cm^2, and a G_c=1 corresponds to a G_m=.2 mmho/cm^2.

changed significantly by different assumptions about the several factors. An increase in the scatterer number density by an order of magnitude will increase the V_m values by a similar amount. An increase in the ionic number density from its value of 10^{-4} molar by an order of magnitude will increase both the current and conductance values by a like amount. An increase in collision frequency will result in an increase in V_m and decrease in the conductances. For example, with ion fixed charge interactions an increase in the fixed charge valence from -1 to -2 increases the collision frequency and V_m by a factor of 4 and decreases the conductances by a like amount.

From Table 13.5 it is quite clear that there are enormous differences in the reuults for various molecular interactions, e.g., in G_m. As a means of comparison, Table 13.6 presents the values of G_c for various interactions with $\tilde{E}=0$. Also shown are the corresponding

TABLE 13.6

p	$G_c(\tilde{E}=0)$	$G_m(\tilde{E}=0)$	ν_o (numbers/sec)
-3	2.44	$1.4x10^{-6}$ (MW=44)	$9.94x10^{13}$ (MW=44 and 1000)
		$3.2x10^{-6}$ (MW=1000)	
-1	1.23	$1.6x10^{-4}$	10^{13}
		$2.2x10^{-4}$	$6.83x10^{12}$
0	1.0	$1.6x10^{-2}$	$8.1 x10^{10}$
		$1.8x10^{-2}$	$6.67x10^{10}$
1/3	.96	$8.4x10^{-2}$	$6.74x10^{10}$
		$9.2x10^{-2}$	$5.66x10^{10}$
1	.92	$1.81x10^{-1}$	$6.52x10^{9}$
		$1.85x10^{-1}$	$6.52x10^{9}$

Table 13.6. The connection between the variation in specific membrane conductances at $V_m=0$, and variations in mean ionic collision frequency for different interactions.

chord conductances for a square centimeter of membrane, and a mean collision frequency, $\nu_0 = \beta\, v^p_{T,Na}$, for each type of interaction. This reveals the sources of variance in the membrane related values. For those molecular interactions where the collision frequency decreases with increasing velocity (e.g., Coulombic and ion-fixed permanent dipole interactions) the mean collision frequency is several orders of magnitude larger than the collision frequencies for the other interactions where $p > 0$. This is due to the long range nature of Coulombic and dipole forces.

Of course, these estimated depend dramatically on the area of the membrane assumed available for ion penetration. For example, taking the cases of fixed charge and fixed permanent dipole scatterers, assume that there are 10 IPR's per square micron, but that 0.1 of the membrane surface area is available for ion penetration. This leads to a scatterer number density of $10^{16}/cm^3$, and the values for the membrane potential in Table 13.5 would be correspondingly multiplied by 1.67×10^{-5}. Also, taking into account the altered current densities, 6.25×10^{15} ions would go through this 0.1 cm^2 every second. For an external ionic concentration of 0.1 molar a bombardment rate of 3.25×10^{23} collisions/second on this area of membrane is expected and the number density of ions interacting with scatterers is approximately 10^{-8} that of the external solution. In Table 13.7 recomputed values of V_m, j_m, and G_m based on the preceeding are presented. It is obvious that for these two types of interactions much more "biological" values of membrane potentials and conductances result.

TABLE 13.7

p	$V_m(\tilde{E}=1)$	$j_m(\tilde{I}=1)$	$G_m(G_c=1)$
-3	250 (MW=44)	.044	.176
	28.4 (MW=1000)	.0111	.391
-1	1.13 (MW=44)	.044	38.9
	.2 (MW=1000)	.0111	55.5

Table 13.7. Model specific membrane potentials, currents, and conductances for a membrane with "many" pores.

CHAPTER 14. THE STEADY STATE MICROSCOPIC MODEL WITH SOLUTION ASYMMETRY

In this chapter I explore the consequences for the microscopic model electrical properties of including a concentration gradient. Steady state conditions are assumed to prevail as before, and the electrolyte bounding each side of the membrane is taken to be identical in composition, but not concentration.

A. Solution of the Kinetic Equations. A steady state situation is again under consideration, so $\dot{f}_0 = \dot{f}_1 = 0$ and $(\partial f_1/\partial x) = 0$. However, $(\partial f_0/\partial x) \neq 0$ and the spatial dependence of f_0 must be determined.

Writing (12.24) and (12.25) with the above restrictions in mind

$$2Z\tilde{E}f_1 = 3V^{p+1}(f_0 + \frac{1}{3V}\frac{\partial f_0}{\partial V}) \tag{14.1}$$

and

$$-V^p f_1 = V\frac{\partial f_0}{\partial X} + Z\tilde{E}\frac{\partial f_0}{\partial V} \tag{14.2}$$

result. Eliminating f_1 between equations 14.1 and 14.2 yields

$$\frac{3}{2}V^{2p}(f_0 + \frac{1}{3V}\frac{\partial f_0}{\partial V}) + \frac{(Z\tilde{E})^2}{V}\frac{\partial f_0}{\partial V} = -Z\tilde{E}\frac{\partial f_0}{\partial X} \tag{14.3}$$

where it is to be understood that $f_0 = f_0(\bar{X},V)$. Generally, E is also a function of $\bar{X}$ and so Poissons equation written in the dimensionless form

$$K^2(d\tilde{E}/d\bar{X}) = ZN$$

$$= 4\pi Z \int_0^{\infty} V^2 f_0(\bar{X},V)\,dV$$

where K is a dimensionless constant, must be integrated to give

$$\tilde{E}(\bar{X}) = (4\pi Z/K^2) \int_0^{\bar{X}}\int_0^{\infty} V^2 f_0(\alpha,V)\,dV d\alpha$$

and then combined with (14.3). When this is done,

$$\frac{3}{2}V^{2p}\left(f_0+\frac{1}{3V}\frac{\partial f_0}{\partial V}\right) + \frac{1}{V}\left(\frac{4\pi Z^2}{K^2}\right)^2 \left\{\int_0^{\bar{X}}\int_0^{\infty} V^2 f_0(\alpha,V)\,dV d\alpha\right\}^2$$

$$= -\frac{4\pi Z^2}{K^2}\frac{\partial f_0}{\partial \bar{X}} \int_0^{\bar{X}}\int_0^{\infty} V^2 f_0(\alpha,V)\,dV d\alpha \qquad (14.5)$$

results. This is a particularly messy non-linear integral-partial differential equation of the mixed Volterra-Fredholm type, and I will not attempt to deal with it.

Instead, prudence dictates an approximation and the assumption I make at this point is that the electric field, $\tilde{E}$, is constant within the membrane. This assumption, which was a consequence of the symmetrical boundary conditions of Chapter 13 and was thoroughly discussed in Chapter 7, allows a simple solution of (14.3) by assuming that $f_0(\bar{X},V) = f_0^{\bar{X}}(\bar{X})\,f_0^V(V)$. Thus if $f_0(\bar{X},V)$ is assumed to be separable I obtain two equations from (14.3) for $f_0^{\bar{X}}(\bar{X})$ and $f_0^V(V)$:

$$(df_0^{\bar{X}}/d\bar{X}) = Af_0^{\bar{X}} \qquad (14.6)$$

and

$$\left[1 + \frac{2(Z\tilde{E})^2}{V^{2p}}\right]\frac{df_0^V}{dV} + \left(3 + \frac{2ZA\tilde{E}}{V^{2p}}\right)Vf_0^V = 0 \qquad (14.7)$$

Thus, the spatially dependent portion of $f_0(\bar{X},V)$ is given by

$$f_0^{\bar{X}}(\bar{X}) = \exp\,(A\bar{X}) \qquad (14.8)$$

If the membrane is of thickness $\bar{\delta}$ and the conditions $N=N_1(N_2)$ and

$\tilde{\varphi} = \tilde{\varphi}_1(\tilde{\varphi}_2)$ at $\bar{X}=0(\bar{\delta})$ hold at the boundaries, then $\tilde{\varphi}(\bar{X}) = \tilde{\varphi}_1 + [(\tilde{\varphi}_1-\tilde{\varphi}_2)/\bar{\delta}]\,\bar{X}$ across the membrane and $\tilde{E} = -\partial\tilde{\varphi}/\partial\bar{X} = -\tilde{\varphi}_m/\bar{\delta}$ where $\tilde{\varphi}_m = \tilde{\varphi}_1-\tilde{\varphi}_2$ is the dimensionless membrane potential. Further

$$\frac{N_2}{N_1} = \frac{\int_0^\infty V^2 f_0(\bar{\delta},V)\,dV}{\int_0^\infty V^2 f_0(0,V)\,dV} = \exp(A\bar{\delta}) \tag{14.9}$$

to require that the separation constant, A, is

$$A = \bar{\delta}^{-1}\ln(N_2/N_1) \tag{14.10}$$

The equation 14.7 for the velocity dependent portion of $f_0(\bar{X},V)$ may be formally integrated to give

$$f_0^V(V) = \exp(-W_0-W_1) \tag{14.11}$$

wherein

$$W_0 = 3\int_0^V \frac{r^{2p+1}dr}{r^{2p} + 2(Z\tilde{E})^2} \tag{14.12}$$

and

$$W_1 = 2ZA\tilde{E}\int_0^V \frac{rdr}{r^{2p} + 2(Z\tilde{E})^2} \tag{14.13}$$

Note that in the absence of a concentration gradient $N_1=N_2$, $A=0$, $W_1=0$, $f_0^X=1$, and f_0^V is identical with the equation (13.13) for f_0 obtained in Chapter 13.

B. A Generalized Goldman Equation. Now that the form of $f_0(\bar{X},V)$ has been determined, an equation for $\tilde{I}(N_1,N_2,\tilde{E})$ may be obtained. First recall that integration of (11.36) yields a macroscopic equation connecting j,E,n, and dn/dx, i.e.

$$j = q^2 n\mu E - qd(nD)/dx \tag{12.13}$$

where μ and D are, respectively, $\overline{\nu^{-1}}$ and $\overline{u\nu^{-1}}$. Rewriting (12.13) in terms of the dimensionless variables of Chapter 12.

$$\tilde{I} = Z^2 N\bar{\mu}\tilde{E} - Zd(N\bar{D})/d\bar{X} \tag{14.14}$$

results wherein $\bar{\mu} = m\nu_0\mu$ is given by

$$\bar{\mu} = - \frac{\int_0^\infty V^{3-p}(\partial f_0/\partial V)dV}{3\int_0^\infty V^2 f_0 dV} \tag{14.15}$$

while $\bar{D} = (\nu_0/V_T^{\,2})D$ is

$$\bar{D} = \frac{\int_0^\infty V^{4-p} f_0 dV}{3\int_0^\infty V^2 f_0 dV} \tag{14.16}$$

Now clearly from my assumption about the separability of $f_0(\bar{X},V)$ $\bar{\mu} \neq \bar{\mu}(\bar{X})$ and $\bar{D} \neq \bar{D}(\bar{X})$. If, further, I now make the constant field assumption then f_0 in equations 14.15-14.16 is identified with (14.11) and

$$\bar{\mu}(A,E) = \frac{3-p}{3} \frac{\int_0^\infty V^{2-p}\exp(-W_0-W_1)dV}{\int_0^\infty V^2\exp(-W_0-W_1)dV} \tag{14.17}$$

and

$$\bar{D}(A,E) = \frac{\int_0^\infty V^{4-p}\exp(-W_0-W_1)dV}{3\int_0^\infty V^2\exp(-W_0-W_1)dV} \tag{14.18}$$

where integration by parts was used to obtain the final expression for $\bar{\mu}$. Thus, A and $\tilde{E}$ both determine $\bar{\mu}$ and $\bar{D}$ and A is to be interpreted as a driving force for ion movement just as $\tilde{E}$ is. The following considerations elaborate this point.

From (14.14), if the electric field is such that $\tilde{I}=0$, $\tilde{E}=\tilde{E}_e$, the equilibrium field, and

$$Z N \bar{\mu} \tilde{E}_e = D(dN/D\bar{X}) \tag{14.19}$$

must be satisfied. Integrating equation 14.19 from $\bar{X} = 0$ to $\bar{X} = \bar{\delta}$ gives

$$Z \bar{\mu} \tilde{E}_e \int_0^{\bar{\delta}} d\bar{X} = \bar{D} \int_{N_1}^{N_2} d \ln N$$

or

$$\begin{aligned} E_e &= \frac{\bar{D}(A, E_e) \ln(N_2/N_1)}{Z \delta \bar{\mu}(A, E_e)} \\ &= A\bar{D}(A, E_e)/Z\bar{\mu}(A, E_e) \end{aligned} \tag{14.20}$$

and thus the equilibrium field $\tilde{E}_e$ is given by the solution of

$$Z \bar{\mu} \tilde{E}_e - A\bar{D} = 0 \tag{14.21}$$

Substituting the expressions given in equations 14.15 and 14.16 for $\bar{\mu}$ and $\bar{D}$ respectively, it is easy to show that (14.21) will be satisfied for all V if and only if

$$(\partial W_0/\partial V) + (\partial W_1/\partial V) = 9\,AV/Z\tilde{E}_e \tag{14.22}$$

Substitution of the explicit forms for W_0 and W_1 into (14.22) then gives $Z\tilde{E}_e = A/3$ and if I convert back to normal (as opposed to dimensionless) variables the Nernst equation results. (Note that the above argument does not establish the Nernst equation, a consequence of very general thermodynamic arguments and not dependent on any assumptions concerning $\tilde{E}$ or f_0. The argument serves only to establish the connection between the separation constant A and the equilibrium field strength).

If the above results are incorporated into (14.12) and (14.13) then I may write

$$W = W_0 + W_1 = 3 \int_0^{\infty} \frac{(r^{2p} + 2Z^2 \tilde{E}\tilde{E}_e)\, r dr}{r^{2p} + 2(Z\tilde{E})^2} \tag{14.23}$$

and it is immediately obvious that as $\tilde{E} \rightarrow \tilde{E}_e$, $W \rightarrow 3V^2/2$ and f_0^V assumes its expected Maxwellian form at equilibrium.

The above considerations with respect to the spatial independence of $\tilde{E}$, $\bar{\mu}$ and $\bar{D}$, and the specified boundary conditions on each side of the membrane allow an easy integration of equation 14.14, yielding

$$\tilde{I} = Z\tilde{E}\bar{\mu}\,\frac{N_1\exp(\bar{\delta}\,Z\tilde{E}\bar{\mu}/\bar{D}) - N_2}{\exp(\bar{\delta}\,Z\tilde{E}\bar{\mu}/\bar{D}) - 1} \qquad (14.24)$$

Equation 14.24 is a generalized form of the Goldman equation of Chapter 7, connecting $\tilde{I}$ with $\tilde{E}$, N_1, and N_2. $\bar{\mu}$ and $\bar{D}$ are to be calculated from (14.17) and (14.18) with equations 14.12 and 14.13 respectively providing the variation of W_0 and W_1 with respect to $\tilde{E}$ and $\tilde{E}_e$. The extension of the Goldman treatment that is embodied in (14.24) is thus to be understood simply as an expression of the dependence of $\bar{\mu}$ and $\bar{D}$ on the two ionic driving forces present in the membrane: $\tilde{E}$ and $\tilde{E}_e$. This dependence disappears if and only if $\tilde{E}=\tilde{E}_e$. Then the system will be at equilibrium with the Einstein relation once again being obeyed.

C. Computed Properties of the Model. The expressions for $\bar{\mu}$ and $\bar{D}$ derived in the previous section are too complicated to be evaluated analytically except in special cases. Thus, a numerical procedure was used to examine the model membrane electrical properties, expressed implicitly in equations 14.17, 14.18, and 14.24. In examining the model three types of ion-membrane molecule interactions, characterized by p=0 and $p = \pm\, 3/4$, were used. This was done to reveal any general characteristic that might emerge for ion membrane molecule collision frequencies that are increasing ($p > 0$) or decreasing ($p < 0$) functions of ionic velocity.

In Figure 14.1 I have plotted the chord conductance $G_c=\tilde{I}/(\tilde{E}-\tilde{E}_e)$, of the model as a function of $\tilde{E}$ for various values of the concentration ratio across the membrane, $\bar{\delta} = 1$, and $p = +3/4$. The figure illustrates that concentration ratios within the physiologically encountered range are capable of producing chord conductance versus electric field strength plots similar to those observed experimentally, e.g. in the membrane of the squid giant axon (Hodgkin and Huxley, 1952).

Though the curves of Figure 14.1 are similar to known data, computations from the Goldman equation yield qualitatively equivalent results. To compare the results of the Goldman and molecular formulations of electrodiffusion theory, define the ratio $R(\tilde{E},\tilde{E}_e)=\tilde{I}_G/\tilde{I}$ where $\tilde{I}_G$ is the Goldman equation current. R as a function of $\tilde{E}$ is

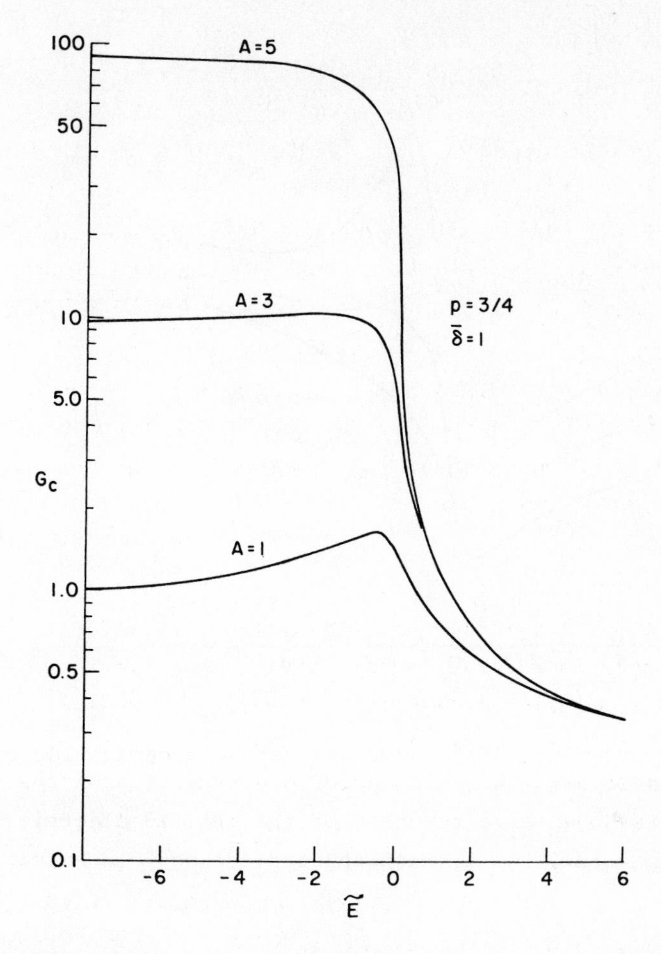

Figure 14.1 Chord conductance as a function of dimensionless electric field strength for $p = 3/4$, $\bar{\delta} = 1$, and three different concentration ratios.

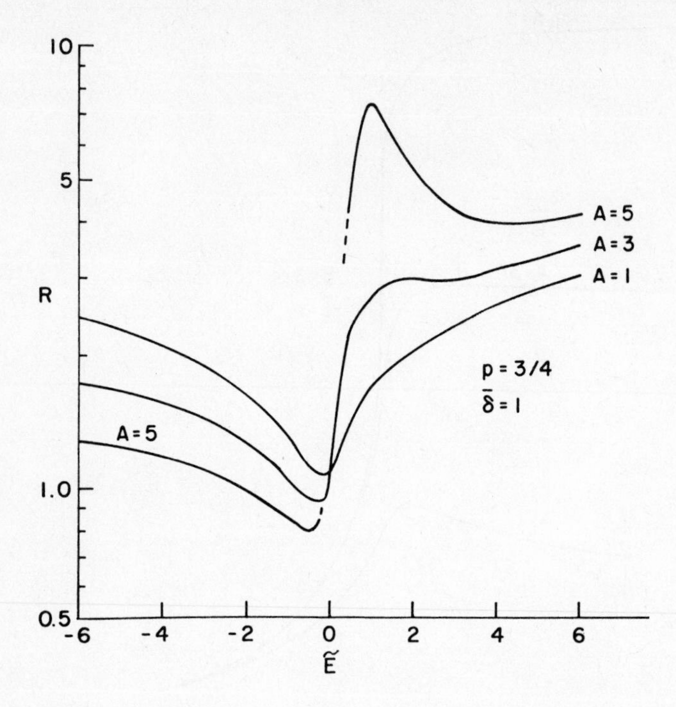

Figure 14.2 A comparison of the current carrying capacities of the microscopic and Goldman formulations of electrodiffusion theory. Here $R = \tilde{I}_G/\tilde{I}$ is shown as a function of the applied electric field strength when $p = 3/4$, $\bar{\delta} = 1$, with three different concentration ratios.

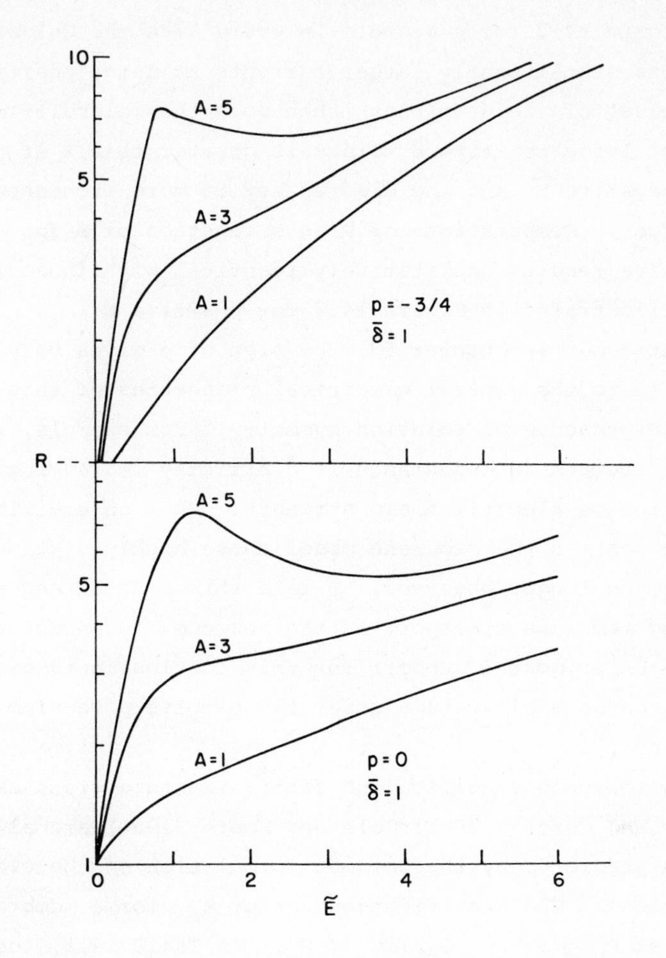

Figure 14.3 As in Figure 14.2, but with $p = -3/4$ and $p = 0$. Note the qualitative similarity in the behaviour of $R(\tilde{E},A)$ for all values of p.

shown in Figure 14.2 for $p = 3/4$. In every case the Goldman equation predicts significantly larger currents at large positive and negative values of field strength than does the molecular model. R for $A > 0$ at large positive $\tilde{E}$ values is greater than R at an equally large but negative $\tilde{E}$, and the discrepancy is more pronounced as A becomes larger. Computations of R as a function of E for $p = -3/4$ and $p = 0$ give results qualitatively identical with those for $p = 3/4$. These are illustrated in Figure 14.3 for positive E.

As pointed out in Chapter 13, the sign of p gives very distinctive features to the general electrical properties of this molecular model in the presence of solution symmetry. For example, with $p > 0 (< 0)$ the chord conductance is a strictly decreasing (increasing) function of electric field strength. When an equilibrium field is present in the membrane model these highly distinctive features are no longer observed. I take this as a strong argument for studying membrane transport in the absence of concentration gradients. Experimental support for this conclusion is given by the studies on several systems under iso-osmotic potassium conditions (c.f. Chapter 5).

Measured current rectification ratios in squid giant axon membrane (Cole and Curtis, 1941; Cole and Moore, 1960) are always larger than predicted by the Goldman formulation of the electrodiffusion model. The rectification ratio, R_R, for a membrane system is defined as $R_R = \lim_{\tilde{E} \to -\infty} G_c / \lim_{\tilde{E} \to +\infty} G_c$. In Table 14.1, for $p=3/4$ and $\bar{\delta} = 1$, the rectification ratios expected for the Goldman equation (equal to N_2/N_1) are compared with the values found from the present formulation. In every case the molecular model gives ratios greater than those expected using the traditional approach and closer to those measured experimentally.

Another point of contention between experimental data and the traditional electrodiffusion development has been the predicted zero-field intersection of the high field asymptotes to the current versus electric field strength curve. Experimentally, a non-zero intersection is found (Cole and Curtis, 1941; Cole and Moore, 1960). The asymptote to the current versus electric field curve will intersect the origin only if the curve has become linear with a slope of $\tilde{I}/\tilde{E}$ at high field strengths. Figures 14.2 and 14.3 both illustrate that this does not occur in the microscopic electro-diffusion model.

TABLE 14.1

	R_R	
A	Goldman	This study
1	2.72	3.02
2	7.39	8.95
3	20.09	29.41
4	54.60	90.04
5	148.41	273.02

Table 14.1. A comparison of the theoretical rectification ratios expected from the Goldman and molecular formulations for p=3/4 and $\bar{\delta}$ = 1. Note the increased rectification efficiency of the molecular model as compared with the Goldman version.

D. One Way Fluxes and the Independence Principle. The differential equation 14.14 connecting $\tilde{I}$ with N, $dN/d\bar{X}$, $\bar{\mu}$, $\bar{D}$, and $\tilde{E}$ is valid irrespective of the constant field assumption used in the previous sections. If I assume that f_0 is separable into velocity and spatially dependent components, then from section A $\bar{\mu}$ and $\bar{D}$ are independent of $\bar{X}$ and (14.14) becomes

$$\tilde{I} = -Z\left[ZN\bar{\mu}(d\tilde{\varphi}/d\bar{X}) + \bar{D}(dN/d\bar{X})\right] \tag{14.25}$$

where I have used $\tilde{E} = -d\tilde{\varphi}/d\bar{X}$. An integrating factor for (14.25) is $\exp(Z\bar{\mu}\tilde{\varphi}/\bar{D})$ and multiplication of (14.25) by this factor along with subsequent integration yields

$$\tilde{I} = Z\,\frac{N_1\exp(Z\bar{\mu}\tilde{\varphi}_1/\bar{D}) - N_2\exp(Z\bar{\mu}\tilde{\varphi}_2/\bar{D})}{\int_0^{\bar{\delta}}\exp(Z\bar{\mu}\tilde{\varphi}/\bar{D})\,d\bar{X}} \tag{14.26}$$

If I define $\tilde{I}_{12}$ as the current flowing from side 1 to side 2 of the membrane, and $\tilde{I}_{21}$ in an analogous fashion then the one way absolute flux ratio is defined as

$$R_F = \tilde{I}_{12}/\tilde{I}_{21} \tag{14.27}$$

From (14.26) this expression is easily obtained, and is

$$R_F = (N_1/N_2)\exp(-Z\bar{\mu}\tilde{\varphi}_m/\bar{D}) \tag{14.28}$$

where I have used $\tilde{\varphi}_m = \tilde{\varphi}_1 - \tilde{\varphi}_2$. Further, from purely thermodynamic considerations, $(N_1/N_2) = \exp(3Z\tilde{\varphi}_e)$ so (14.28) takes the form

$$R_F = \exp\left[-Z\left(\frac{\bar{\mu}\tilde{\varphi}_m}{\bar{D}} - 3\tilde{\varphi}_e\right)\right] \tag{14.29}$$

For a classical electrodiffusion model, $(\bar{\mu}/\bar{D}) = 3$, and under these conditions equation 14.29 becomes

$$R_{F,ED} = \exp\left[-3Z(\tilde{\varphi}_m - \tilde{\varphi}_e)\right] \tag{14.30}$$

which is often referred to as the Ussing flux ratio from its original derivation (Ussing, 1949).

In a classical study Hodgkin and Keynes (1955) examined one-way ionic fluxes through the potassium channel of the squid giant axon membrane with radioactive tracers. They noted that the experimental flux ratios did not fit the form expected from the Ussing equation 14.30, but rather that they were empirically described by an equation of the form

$$R_F = \exp\left[-Zn(\tilde{\varphi} - \tilde{\varphi}_e)\right] \tag{14.30a}$$

where $n \sim 3$ ($n=1$ from electrodiffusion theory). The Ussing relation, obtained by integrating the Nernst-Planck equation, does not explicitly consider the effects of ion-ion interaction. The discrepancy between the predictions of the Ussing relation and experimental findings has been widely interpreted as evidence for the interaction of ions with each other as they cross the membrane, and as evidence against the importance of electrodiffusion mechanisms in determining movement through the potassium channel. However, it is clear that based on considerations from the molecular formulation of electrodiffusion significant deviations from the classical Ussing relation are not unexpected and further that these deviations are in no way connected with ion-ion interaction. Thus it is of some interest to examine the flux ratio theoretically predicted from equation 14.29. To facilitate comparison with earlier computations I take the constant field assumption to be valid so (14.29) becomes

$$R_F = \exp\left[Z\bar{\delta}\left(\frac{\bar{\mu}\tilde{E}}{D} - 3\tilde{E}_e\right)\right] \tag{14.31}$$

and the classical equation, (14.30), is

$$R_{F,ED} = \exp\left[3Z\bar{\delta}\,(\tilde{E} - \tilde{E}_e)\right] \tag{14.32}$$

In Figure 14.4, I show the flux ratio as a function of $(\tilde{E}-\tilde{E}_e)$ for $p=3/4$, $\bar{\delta}=1$, and three different values of A. For comparison the classical result, given by equation 14.32 , is shown as a dotted line. The predictions of the molecular electrodiffusion model are quite different from those of classical electrodiffusion theory. For small departures of $(\tilde{E}-\tilde{E}_e)$ from zero , the computed results may be fit by an equation of the form of (14.30a), with $n < 1$. This is opposed to the result obtained by Hodgkin and Keynes, but is obviously dependent only on $\bar{\delta}$.

In Figures 14.5 and 14.6 the theoretical flux ratios are given as functions of $(\tilde{E}-\tilde{E}_e)$, with $\bar{\delta}=1$, several values of A, and $p= -3/4$ and $p = 0$ respectively. The curves are generally similar to those found with $p = 3/4$. The comments about fitting the theoretical curves with equation 14.30a apply again.

The computed results presented above show that flux ratios obtained from an electrodiffusion model of ion transport need not have the form predicted by equation 14.32 if interaction between ions is negligible. The molecular model shows behavior departing significantly from the theoretical expectations of the classical formulation. Yet, as pointed out above, ion-ion interactions are excluded.

What is the significance of the fact that, with $\bar{\delta}=1$, all of the flux ratio versus $(\tilde{E}-\tilde{E}_e)$ curves computed for our model have a slope about $\tilde{E}=\tilde{E}_e$ that is less than the classical prediction? That is, the curves in the neighborhood of $\tilde{E}=\tilde{E}_e$ can be fit by equation 14.30a with $n < 1$. In Figure 14.7 I illustrate the effect of changing $\bar{\delta}$ on the flux ratio versus $(\tilde{E}-\tilde{E}_e)$ curves for $p = 3/4$ and $(N_2/N_1) = 20.09$ (corresponding to $A=3$ when $\bar{\delta}=1$). By varying the value of $\bar{\delta}$ it is possible to overcome the discrepancy noted above and produce curves that, around $\tilde{E}=\tilde{E}_e$, can be fit by (14.31) with $n > 1$. Similar results are found for all values of p, but are not reproduced.

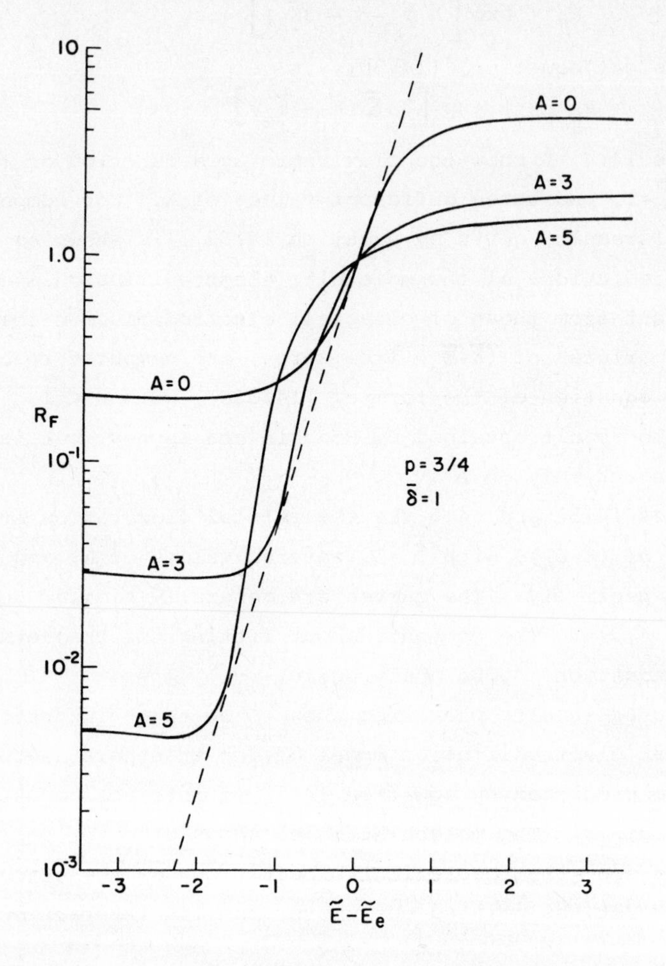

Figure 14.4 The flux ratio, R_F, as a function of the ionic driving force, $\tilde{E}-\tilde{E}_e$, across the membrane for $\bar{\delta} = 1$ and $p = 3/4$. The computations are for no concentration gradient (A = 0) and concentration ratios of 20.1 (A = 3) and 148 (A = 5). The dashed line is the classical result predicted by equation 14.32.

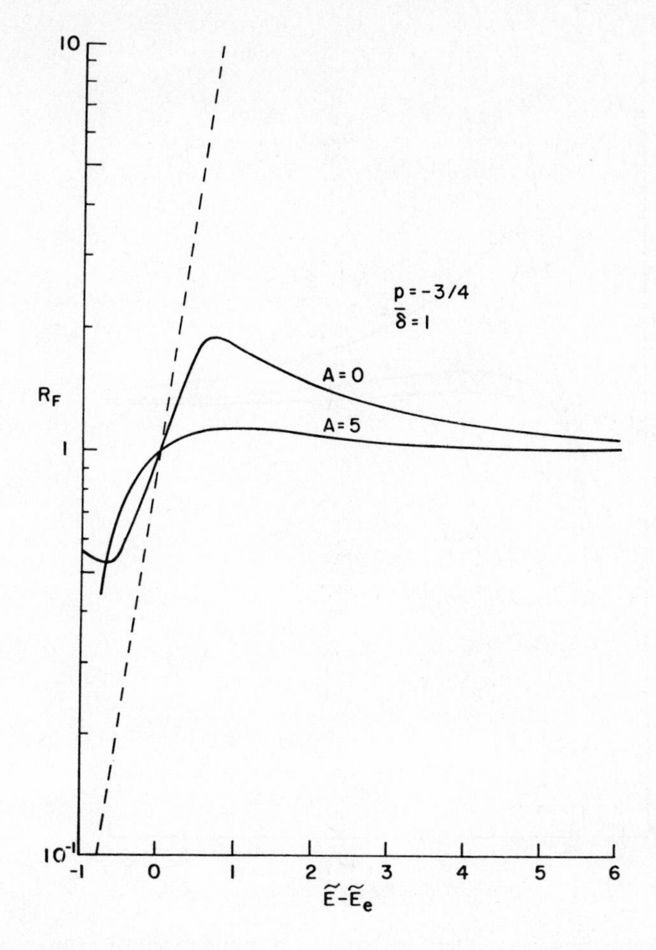

Figure 14.5 As in Figure 14.4, but for $p = -3/4$ and two values of the concentration ratio.

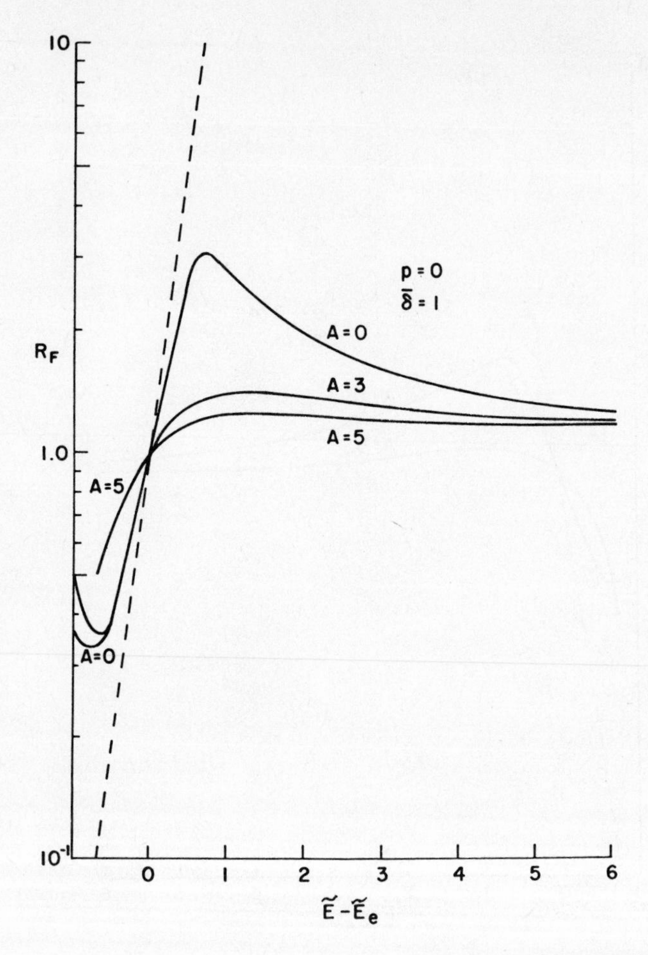

Figure 14.6 The one way flux ratio as a function of ionic driving force for p = 0, $\bar{\delta}$ = 1, and three concentration ratios.

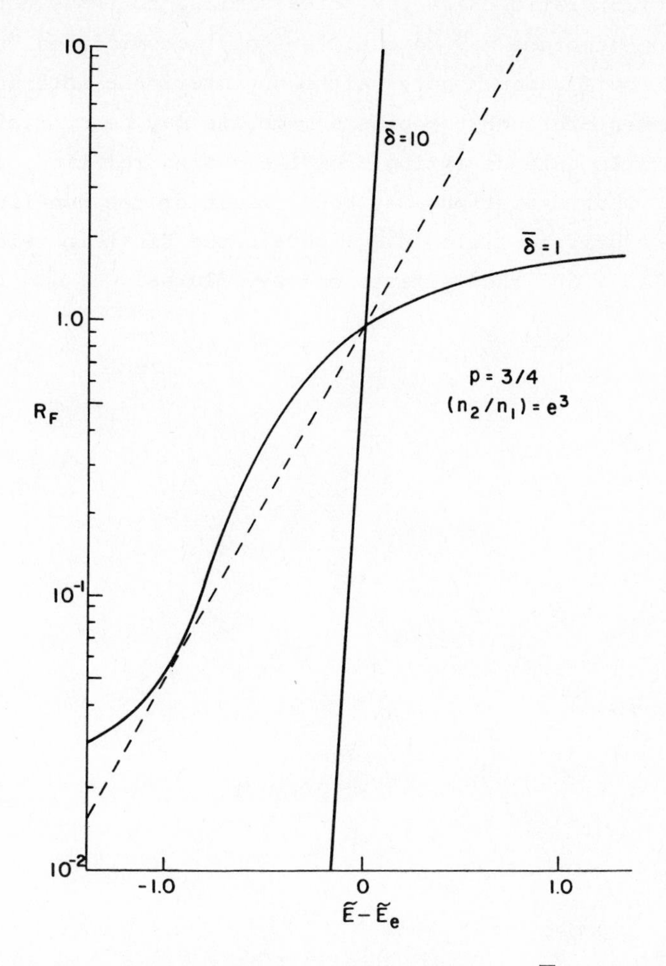

Figure 14.7 The influence of the parameter $\bar{\delta}$ on computed R_F versus $\tilde{E}-\tilde{E}_e$ curves with a fixed concentration ratio of 20.1. Alterations of $\bar{\delta}$ rotate the curve about the origin, as expected and are capable of qualitatively matching experimentally observed behaviour.

Therefore, flux-ratio characteristics similar to those noted in squid giant axon membrane may be qualitatively accommodated by the molecular electrodiffusion model. Although interionic interaction during ion movement through the plasma membrane may be sufficient to yield flux ratio data deviating from the Ussing relation, it is not necessary. Such deviations may be a result of the non-linear way in which the driving field ($\tilde{E}-\tilde{E}_e$) alters the Einstein relation and ionic mobility, and thus affects one-way fluxes.

CHAPTER 15. STEADY STATE AND DYNAMICAL PROPERTIES OF THE MACROSCOPIC MODEL

In Chapter 6 I developed a macroscopic formulation of electrodiffusion theory that was used in Chapters 7 through 10 to examine transmembrane ion movement when $t_c \ll t_D$ and ionic energy is essentially unaltered by the presence of electric fields or concentration gradients. In Chapter 12 I connected the macroscopic model with the microscopic model developed in Chapter 11. In this chapter I examine some of the steady state and time dependent properties of the macroscopic model.

Instead of the dimensionless variables of Chapter 6, here I will use the alternate set introduced at the end of Chapter 12. With these definitions, equations 6.7 and 6.8 become

$$\frac{1}{N}\frac{\partial(NV)}{\partial T} = \tilde{E} - VU^{p/2} - \frac{1}{3N}\frac{\partial(NU)}{\partial X} \tag{15.1}$$

and

$$\frac{1}{N}\frac{\partial(NU)}{\partial T} = 2V\tilde{E} - \frac{1}{N}\frac{\partial(NUV)}{\partial X} + \xi(1-U)U^{p/2} \tag{15.2}$$

respectively, and (6.11) becomes

$$I = ZNV \tag{15.3}$$

If, in the steady state, a concentration gradient exists across the membrane, equations 15.1 through 15.3 become

$$I = \frac{ZN\tilde{E}}{U^{p/2}} - \frac{Z}{3U^{p/2}}\frac{d(NU)}{dX} \tag{15.4}$$

and

$$0 = 2I\tilde{E} - \frac{I}{N}\frac{d(NU)}{dX} + \xi ZN(1-U)U^{p/2} \tag{15.5}$$

I assume $\tilde{E}$ to be independent of X, which implies that U and ν are also independent of X (c.f. Chapter 14). If the membrane is of thickness $\bar{\delta}$, and if at X=0 ($\bar{\delta}$) the conditions $N=N_1(N_2)$ and $\varphi = \varphi_1(\varphi_2)$ hold, then $\varphi(X) = \varphi_1 + (\varphi_2-\varphi_1)X/\bar{\delta}$ and $E = -d\varphi/d = -\varphi_m/\bar{\delta}$, where $\varphi_m = \varphi_1-\varphi_2$ is the membrane potential.

With the above assumptions and boundary conditions, equation 15.4 is easily integrated to give

$$I = \frac{Z\tilde{E}}{U^{p/2}} \frac{N_2 - N_1\exp(3\bar{\delta}\tilde{E}/U)}{1 - \exp(3\bar{\delta}\tilde{E}/U)} \tag{15.6}$$

The expression for the current density requires knowledge of how U depends on $\tilde{E}$, N_1, and N_2. I obtain this by integrating equation 15.5 again obtaining an expression for I, and equating the result with (15.6). The result is the trancendental equation

$$\frac{\xi U^p(U-1)}{\tilde{E}^2} = \frac{2N_2}{N_1}\left[\frac{N_2 - N_1\exp(3\bar{\delta}\tilde{E}/U)}{1 - \exp(3\bar{\delta}\tilde{E}/U)}\right] \tag{15.7}$$

which implicitly gives $U = U(\tilde{E},N_1,N_2)$.

From equation 15.6 as $(\tilde{E}/U) \rightarrow -(3\bar{\delta})^{-1}\ln(N_1/N_2)$, $I \rightarrow 0$. I define the value of $\tilde{E}$ such that $I = 0$ as the equilibrium field $\tilde{E}_e$ and the corresponding membrane potential, the equilibrium potential, φ_e. From (15.7), as $I \rightarrow 0$, $U \rightarrow 1$, and the equilibrium field is given by

$$\tilde{E}_e = -(3\bar{\delta})^{-1}\ln(N_1/N_2) \tag{15.8}$$

Use of the equilibrium field defined by equation 15.8 allows us to write (15.6) and (15.7) in a more symmetric form; namely

$$I = \frac{\tilde{E}\sqrt{N_1/N_2}}{U^{p/2}} \frac{\sin h\left[\bar{\delta}(\tilde{E}-U\tilde{E}_e)/U\right]}{\sin h\left[\bar{\delta}\tilde{E}/U\right]}, \tag{15.9}$$

and

$$\xi U^p(U-1) = 2\tilde{E}^2 \frac{\sin h^2\left[\bar{\delta}(\tilde{E}-U\tilde{E}_e)/U\right]}{\sin h^2\left[\bar{\delta}\tilde{E}/U\right]} \tag{15.10}$$

respectively.

The generalized Goldman equation derived above, equation 15.9, in conjunction with (15.10), has behavior virtually identical to that described in Chapter 13 and 14 for the microscopic model. Comparison between the two analyses indicates that $U^{-p/2}$ plays the role of a dimensionless mobility, $\bar{\mu}$, while $3^{-1}U^{(2-p)/2}$ is a dimensionless diffusion coefficient, $\bar{D}$. The advantage of the macroscopic model analysis is that U may be obtained as the solution of a trancendental equation. In the microscopic model, the numerical computation of three integrals was required to determine $\bar{\mu}(\tilde{E},\tilde{E}_e)$ and $\bar{D}(\tilde{E},\tilde{E}_e)$.

It should be noted that in a steady state situation with no concentration gradient across the membrane, equations 15.4 and 15.9 become

$$I = Z\tilde{E}U^{-p/2} \tag{15.11}$$

and

$$\xi U^p(U-1) = 2\tilde{E}^2 \tag{15.12}$$

For small field strengths, $(2\tilde{E}^2/\xi) \ll 1$, $U \simeq 1$ (the ionic energy has not been increased significantly over its thermal energy) and the chord conductance $G_c = U^{-p/2}$ is approximately constant. For large field strengths, $(2\tilde{E}^2/\xi) \gg 1$, $U \gg 1$ so (15.12) may be solved approximately to give $U \simeq (2\tilde{E}^2/\xi)^{1/(p+1)}$. Thus for high field conditions the conductance is given by $G_c \simeq (2\tilde{E}^2/\xi)^{-p/2(p+1)}$ and G_c is a decreasing (increasing) function of electric field strength for $p > 0$ $(-1 < p < 0)$. This is the same conclusion reached in the microscopic formulation.

For $p < -1$ the situation is not as simple, for U is an increasing function of $\tilde{E}$ only to a certain value of $\tilde{E}$. To determine for what values of $\tilde{E}$ the dimensionless energy is defined, I must examine the behavior of the equation $(2\tilde{E}^2/\xi) = (U-1)/U^m$, where $m = -p > 1$.

Let $f(U) = (U-1)/U^m$, so it is a simple matter to show that $(df/dU) > 0$ for $1 \leq U < m/(m-1)$. At $U = m/(m-1)$, $(df/dU) = 0$ and $f = m^{-1}\left[(m-1)/m\right]^{m-1}$. For any given m value our formulae are applicable for

$$0 \leq |\tilde{E}| < \left[\frac{\xi}{2m}\left(\frac{m-1}{m}\right)^{m-1}\right]^{1/2}, \quad m > 1$$

For $p = -1(m=1)$ the formulae are applicable for $\tilde{E} < \sqrt{\xi/2}$.

Two classes of classical interactions are characterized by $p < 0$ for ion-permanent dipole collisions $p = -1(m=1)$, and for ion-fixed charge (coulombic) collisions $p = -3(m=3)$. For these two classes of interactions, the normalizing constants relating dimensionless $\tilde{E}$ values to actual membrane potentials may be calculated from first principles. When this is done the $\tilde{E}_{max}$ values for which our formulae apply, i.e. $(\tilde{E}_{max}/\sqrt{\xi}) = .707$ for $p = -1$, $(\tilde{E}_{max}/\sqrt{\xi}) = .273$

for p = -3, corresponds to membrane potentials much greater than those likely to be encountered in biological situations (c.f. Table 13.7).

In this section I examine the time course of the model conductance changes in response to an "instaneously" applied electric field. Valuable insight into the mechanisms of these changes is provided by examining the response of the system in the absence of concentration gradients, so I confine my attention to this situation.

In the absence of a concentration gradient, the model equations become

$$\frac{dV}{dT} = \tilde{E} - VU^{p/2} \tag{15.13}$$

and

$$\frac{dU}{dT} = 2V\tilde{E} + \xi (1-U)U^{p/2} \tag{15.14}$$

respectively. This system has no simple closed solution except for p=0.

If a field, $\tilde{E}$, is suddenly applied across the membrane, the initial effect will be to accelerate the ions so $\dot{V} \neq 0$. For short times ($t \ll \nu_0^{-1}$ or $T \ll 1$) there will be little change in ionic energy, as this is accomplished through collisional energy losses. Hence, for $T \ll 1$ I expect $\dot{U} \simeq 0$. If there is no field across the membrane before $\tilde{E}$ is applied, $U \simeq 1$ and the early time behavior will be approximately described by $\dot{V} + V = E$. Thus

$$G(T) = (I/\tilde{E}) = 1 - \exp(-T) \tag{15.15}$$

In Figure 15.1 the full behavior of G(T), computed from equations 15.13 and 15.14, is shown for a range of $\tilde{E}$ and p = 1/3, 1/2, 1, and 4. (p=1/3 corresponds to a classical induced-dipole collision; p=1 characterizes an ion-neutral particle interaction, c.f. Chapter 11). At early times ($T \ll 1$), when accelerative effects are expected to dominate, G(T)is indeed closely approximated by (15.15) which is shown as a dashed line for each value of p. The deviation from the behavior predicted by equation 15.15 will arise as collision-induced ionic energy losses become appreciable. The behavior in Figure 15.1 illustrates that increases in p and/or $\tilde{E}$ enhance the early appearance of these phenomena, as would be expected.

The numerical solutions of (15.13) and (15.14) shown in Figure

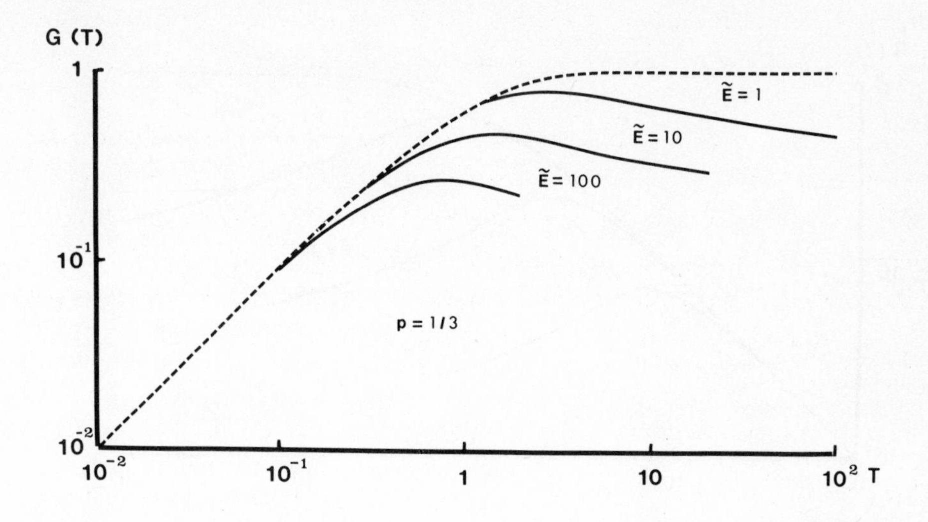

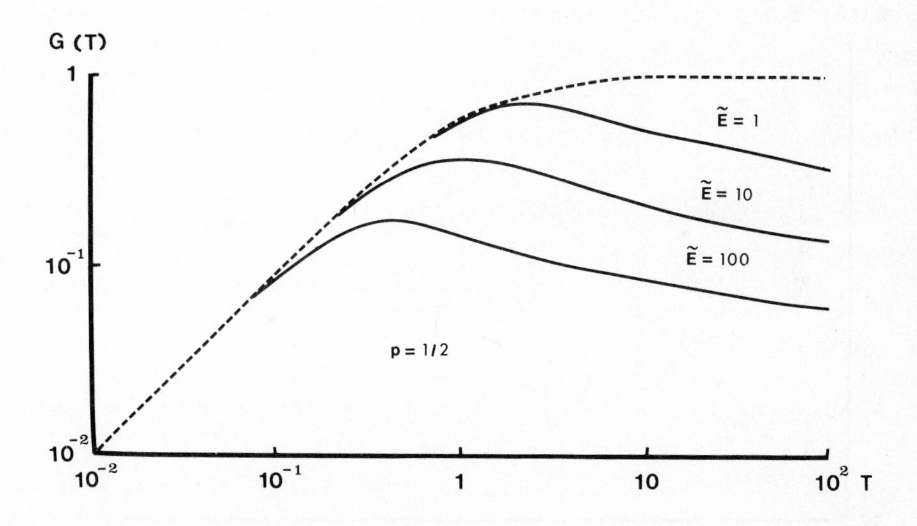

Figure 15.1 Full logarithmic plots of G(T) as a function of T in response to a step change in $\tilde{E}$ from zero to the indicated value at T = 0, as obtained from the numerical solutions of equations 15.13 and 15.14. The computations were carried out for p = 1/3, 1/2, and (next page) 1 and 4 with $\zeta = 10^{-3}$ in every case. The dashed line gives the response for p = 0, G(T) = 1 - exp(-T), which is also the approximate early time behaviour according to (15.15).

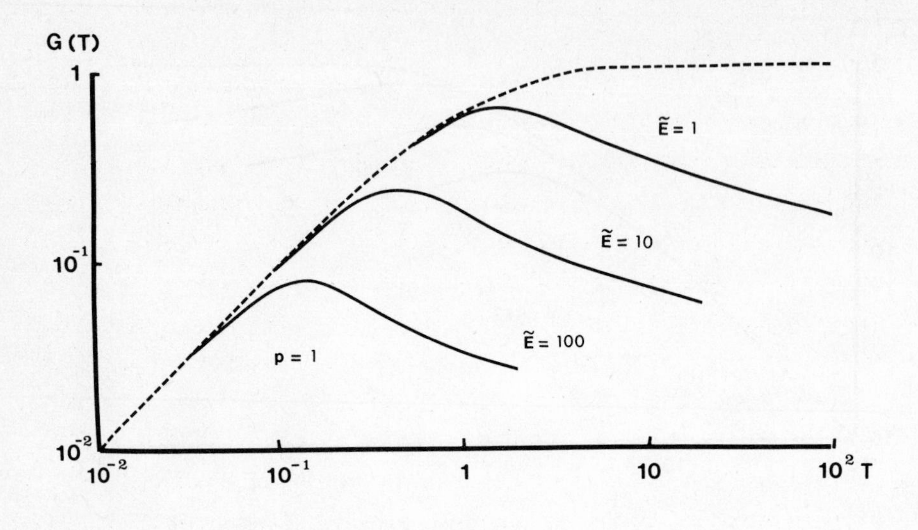

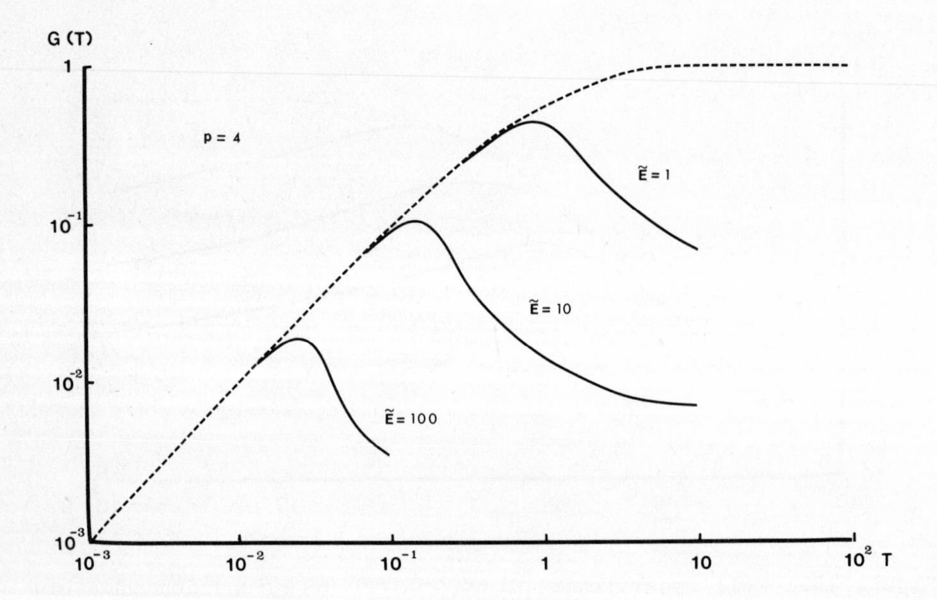

Figure 15.1 continued

15.1 indicate that G(T) reaches a maximum, then declines to a final steady state value. If I denote the maximum value of G(T), reached at time T_m, as G_m then both G_m and T_m are decreasing functions of p and E as indicated in Figures 15.2 and 15.3 respectively.

Within the context of the model presented here, the maximum in the G(T) versus T relation simply implies that a point ($\dot{V}=0$) has been reached where the accelerating effect of the electric field on the ion is exactly balanced by the collision-induced deceleration Thus, at the maximum, the net force on the ion is zero even though the ionic energy is still changing. As the trade-off between electric field and collision-induced energy changes continues, the ion exhibits net deceleration ($\dot{V} < 0$) characterized by $\dot{U} > 0$ until a steady state with $\dot{V} = \dot{U} = 0$ is obtained. The dependence of this steady state on the electric field strength is given by the solutions of equations 15.11 and 15.12 above.

In real (as opposed to dimensionless) time the ionic acceleration due to the applied electric field has a characteristic time of ν_0^{-1}, while the ionic deceleration due to collisions has a characteristic time of $(\xi \nu_0)^{-1}$. In Figures 15.1 through 15.5 I used $\xi = 10^{-3}$ in the numerical computations, assuring the separation of these two characteristic times by a factor of 10^3 and the resultant appearance of the maximum in G(T) with respect to time.

To obtain an approximate quantitative picture of the above events, proceed as follows. For $T >> 1$ ($t >> \nu_0^{-1}$), it is expected that the primary source of time variation arises from energy changes. Therefore, approximate the model with $\dot{V} \simeq 0$ and $\dot{U} \neq 0$, so

$$V \simeq \tilde{E}U^{-p/2} \tag{15.16}$$

and

$$\frac{dU}{dT} = 2V\tilde{E} + \xi\,(1-U)U^{p/2} \tag{15.17}$$

If (15.16) and (15.17) are combined the result is

$$U^{p/2}\,\frac{dU}{dT} = 2\tilde{E}^2 + \xi\,(1-U)U^{p/2} \tag{15.18}$$

If in (15.18) I set $U^{p/2} = R$ and take R to be approximately independent of T, I obtain $U(T) \simeq 1 + (2E^2/\xi R)\left[1-\exp(-\xi RT)\right]$. Now R will range from 1 to $U(T \rightarrow \infty)^{p/2}$, so I can get an approximate idea of the extremes of behavior exhibited by the system.

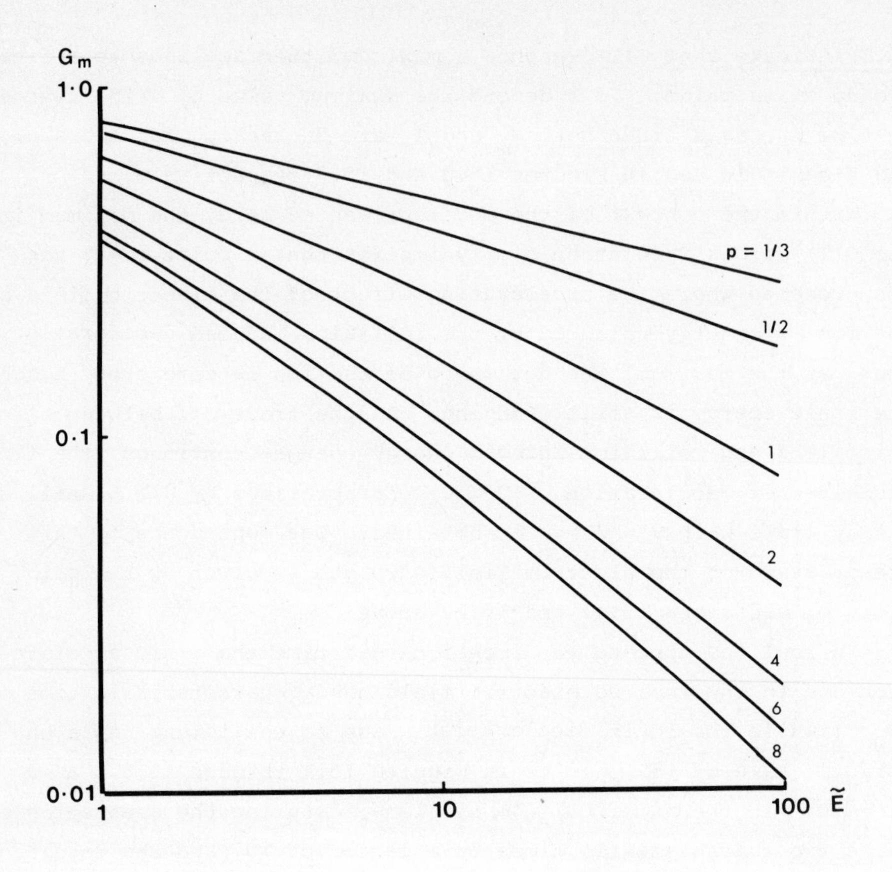

Figure 15.2 The variation of the maximum conductance, G_m, with respect to the applied electric field strength, E, for a range of $p > 0$. $\xi = 10^{-3}$ throughout.

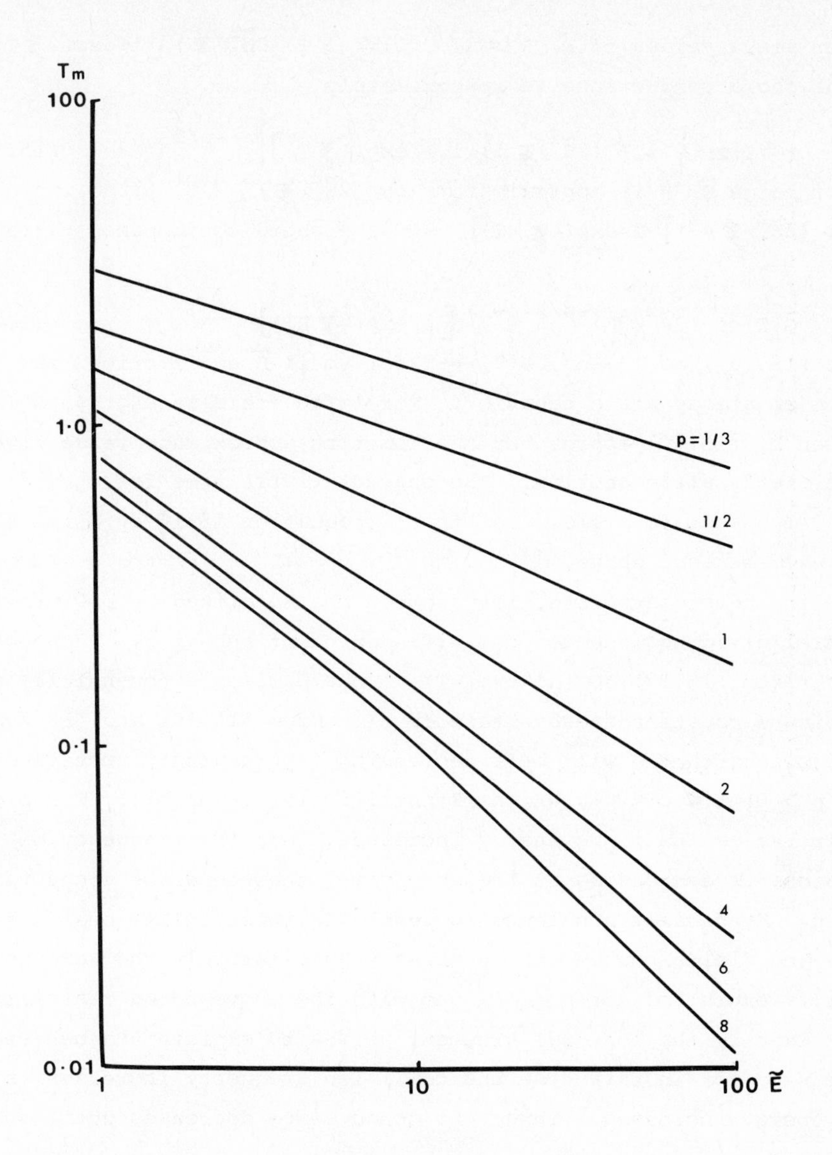

Figure 15.3 The effect of the applied electric field strength $\tilde{E}$ on the time to maximum conductance, T_m, in the model when $p>0$. $\zeta = 10^{-3}$ as before.

For small values of $\tilde{E}$, $R \simeq 1$, $U(T) \simeq 1 + (2\tilde{E}^2/\xi)[1-\exp(-\xi T)]$ and the chord conductance is approximately

$$G(T) \simeq \left\{ 1 + (2\tilde{E}^2/\xi)\left[1 - \exp(-\xi T)\right]\right\}^{-p/2} \tag{15.19}$$

For large $\tilde{E}$, R is approximated by $(2\tilde{E}^2/\xi)^{p/2(p+1)}$, so $U(T) \simeq (2\tilde{E}^2/\xi R^2)[1-\exp(-\xi RT)]$ and the chord conductance varies as

$$\bar{G}(T) \simeq (2\tilde{E}^2/\xi)^{-p/2(p+1)}\left[1-\exp(-\xi RT)\right]^{-p/2} \tag{15.20}$$

In (15.19), as $T \to \infty$, $G(T) \to 1$ for small $\tilde{E}$ as described in the section on steady state behavior. For large field strengths, G(T) as given by (15.20) approaches the limiting approximate value given in the steady state section. The characteristic time for changes in G(T) is also seen to differ from that of equation 15.15 by a factor of ξ as discussed above.

As in the steady state, interaction characterized by p=0 provide a natural dividing line for the time-dependent behavior. When p=0 I have, from (15.13), G = 1-exp(-T). Thus G rises exponentially and maintains a constant steady-state level. From (15.19) and (15.20) it is obvious that G will be a decreasing (increasing) function of T for $p > 0$ $(-1 < p < 0)$, which is intuitively reasonable. For $p < 0$ as T increases the ionic energy increases. But the frequency of collisions is decreasing at the same time, therefore the conductance goes up. Eventually a balance is reached between collisional energy losses and field-induced energy gains. Qualitatively the same behavior is exhibited for $p < -1$, but with the above noted restrictions on $\tilde{E}$. Exactly the converse argument serves to explain the behavior when $p > 0$, for in this case the collision frequency increases as ionic energy increases. Thus, the conductance decreases until a balance is again attained.

All of the computed solutions for equations 15.13 and 15.14 presented here have been for $p > 0$. When p = -1, -2, or -3, and $\xi = 10^{-3}$ the solutions are well described by (15.15) for $T \leq 10$. With this value of ξ there is a very wide separation in time between electric field and collision-related phenomena, and it is more pronounced than with $p > 0$. In any case, the behavior of G(T) in response to a step change in $\tilde{E}$ will initially follow the dashed curve as shown in Figure 15.1, remain at a constant value (=1) for some time, then rise to a second steady state value.

The time-dependent behavior exhibited by the conductance of this model when $p > 0$ is strikingly similar to the known behavior of the sodium conductance in a number of excitable systems (c.f. Chapter 4). In response to a step change in the applied electric field, the conductance rises to a maximum and then decays to a lower steady state value. Both the maximum conductance and the time to maximum vary with respect to the membrane potential, and this variation is at least qualitatively similar to that found experimentally.

Many interpretations of sodium conductance data have started with the assumption that the activation-inactivation sequence is the result of two separate mechanisms operating independently (Hodgkin and Huxley, 1952). The activation-inactivation pattern displayed by the conductance in this model is not a result of the action of two completely separate mechanisms. Rather, it is due to the subtle interplay of two interdependent physical processes, and inactivation of the conductance is a process intimately tied to activation. Indeed, Hoyt (1963,1968) and Hoyt and Adelman (1971) have been able to fit voltage clamp data on g_{Na} using a model in which activation-inactivation is a coupled process, and explain the results of certain inactivation experiments not resolved by the Hodgkin-Huxley formulation of the process.

The conceptual similarity between the Hoyt model and the one I am considering, in conjunction with the results of the two models, might seem to imply that the molecular mechanisms I consider play a significant role in the determination of time-dependent transport processes in excitable membranes. However, this interpretation seems to fail on at least three points.

First, note that the characteristic time (ν_0^{-1}) of this model differs from that found experimentally by at least three orders of magnitude (c.f. Chapter 13). Second, the effect of temperature on the time parameters of this model may be shown to be quite small ($Q_{10} \simeq 1.1$) with respect to those found experimentally ($Q_{10} \simeq 2$-3). The temperature dependence of the steady state model conductances is more in accord with those found in excitable systems. Finally, this model would predict a sensitivity of the time parameters to ionic properties almost as great as those observed in the steady state conductance. Experimental data (Binstock and Lecar, 1969; Chandler and Meves, 1965; Meves and Chandler, 1965; Moore et al., 1966), indicate no change in the time constants of sodium conductance activation and inactivation when current through the sodium channel is carried by a number of monovalent sodium ion substitutes.

REFERENCES

Adelman, W.J. and J.P. Senft. (1966). Voltage clamp studies on the effect of internal cesium ion on sodium and potassium currents in the squid giant axon. J. Gen. Physiol. 50: 279.

Adelman, W.J. and R.E. Taylor. (1961). Leakage current rectification in the squid giant axon. Nature, 190: 883.

Adrian, R.H. and C.L. Slayman. (1966). Membrane potential and conductance during transport of sodium, potassium and rubidium in frog muscle. J. Physiol. 184: 970.

Agin, D. (1963). Some comments on the Hodgkin-Huxley equations. J. Theoret. Biol. 5: 161.

Albuquerque, E.X. and S. Thesleff. (1967). Influence of phospholipase C on some electrical properties of the skeletal muscle membrane. J. Physiol. 190: 123.

Armstrong, C.M. (1966). Time course of TEA^+-induced anomalous rectification in squid giant axons. J. Gen. Physiol. 50: 491.

Armstrong, C.M. and L. Binstock. (1965). Anomalous rectification in the squid giant axon injected with tetraethylammonium chloride. J. Gen. Physiol. 48: 859.

Baker, P.F., A.L. Hodgkin and H. Meves. (1964). The effect of diluting the internal solution on the electrical properties of a perfused giant axon. J. Physiol. 170: 541.

Baker, P.F., A.L. Hodgkin and T.I. Shaw. (1962). Replacement of the axoplasm of giant nerve fibres with artificial solutions. J. Physiol. 164: 330.

Baker, P.F., A.L. Hodgkin and T.I. Shaw. (1962). The effects of changes in internal ionic concentrations on the electrical properties of perfused giant axons. J. Physiol. 164: 355.

Bernstein, J. (1902). Untersuchungen zur Thermodynamik der bioelektrischen Ströme. Pflug. Arch. ges. Physiol. 92: 521.

Binstock, L. and H. Lecar. (1969). Ammonium ion currents in the squid giant axon. J. Gen. Physiol. 53: 342.

Blaustein, M.P. and D.E. Goldman. (1966). Competitive action of calcium and procaine on lobster axon. J. Gen. Physiol. 49: 1043.

Brinley, F.J. and L.J. Mullins. (1967). Sodium extrusion by internally dialyzed squid axons. J. Gen. Physiol. 50: 2303.

Brinley, F.J. and L.J. Mullins. (1974). Effects of membrane potential on sodium and potassium fluxes in squid axons. Ann. N.Y. Acad. Sci. 242: 406.

Butz, E.G. and J.D. Cowan. (1974). Transient potentials in dendritic systems of arbitrary geometry. Biophys. J. 14: 661.

Caldwell, P.C. (1960). The phosphorus metabolism of squid axons. and its relationship to the active transport of sodium. J. Physiol. 152: 545.

Caldwell, P.C., A.L. Hodgkin, R.D. Keynes and T.I. Shaw. (1960). The effects of injecting "energy-rich" phosphate compounds on the active transport of ions in the giant axons of Loligo. J. Physiol. 152: 561.

Caldwell, P.C., A.L. Hodgkin, R.D. Keynes and T.I. Shaw. (1960). Partial inhibition of the active transport of cations in the giant axons of Loligo. J. Physiol. 152: 591.

Caldwell, P.C. and R.D. Keynes. (1959). The effect of ouabain on the efflux of sodium from a squid giant axon. J. Physiol. 148: 8p.

Camougis, G., H. Takman and J.P. Tasse. (1967). Potency difference between the zwitterion form and the cation forms of tetrodotoxin. Science, 156: 1625.

Chandler, W.K., R. Fitzhugh and K.S. Cole. (1962). Theoretical stability properties of a space-clamped axon. Biophys. J. 2: 105.

Chandler, W.K. and H. Meves. (1964). Voltage-clamp experiments on perfused giant axons. J. Physiol. 173: 31p.

Chandler, W.K. and H. Meves. (1965). Voltage clamp experiments on internally perfused giant axons. J. Physiol. 180: 788.

Chandler, W.K., A.L. Hodgkin and H. Meves. (1965). The effect of changing the internal solution on sodium inactivation and related phenomena in giant axons. J. Physiol. 180: 821.

Chapman, S. and T.G. Cowling. (1958). Mathematical theory of non-uniform gases. London: Cambridge University Press.

Cohen, H. and J.W. Cooley. (1965). The numerical solution of the time dependent Nernst-Planck equations. Biophys. J. 5: 145.

Cole, K.S. (1941). Rectification and inductance in the squid giant axon. J. Gen. Physiol. 25: 29.

Cole, K.S. (1949). Dynamic electrical characteristics of the squid axon membrane. Arch. Sci. Physiol. 3: 253.

Cole, K.S. (1965). Electrodiffusion models for the membrane of squid giant axon. Physiol. Rev. 45: 340.

Cole, K.S. (1968). Membranes, ions and impulses. Berkeley: University of California Press.

Cole, K.S. and R.F. Baker. (1941). Transverse impedance of the squid giant axon during current flow. J. Gen. Physiol. 24: 535.

Cole, K.S. and R.F. Baker. (1941). Longitudinal impedance of the squid giant axon. J. Gen. Physiol. 24: 771.

Cole, K.S. and R.H. Cole. (1941). Dispersion and absorption in dielectrics: I. Alternating current characteristics. J. Chem. Phys. 9: 341.

Cole, K.S. and H.J. Curtis. (1938). Electric impedance of Nitella during activity. J. Gen. Physiol. 22: 37.

Cole, K.S. and H.J. Curtis. (1939). Electric impedance of the squid giant axon during activity. J. Gen. Physiol. 22: 649.

Cole, K.S. and H.J. Curtis. (1941). Membrane potential of the squid giant axon during current flow. J. Gen. Physiol. 24: 551.

Cole, K.A. and A.L. Hodgkin. (1939). Membrane and protoplasm resistance in the squid giant axon. J. Gen. Physiol. 22: 671.

Cole, K.S. and J.W. Moore. (1960). Potassium ion current in the squid giant axon: Dynamic characteristics. Biophys. J. 1: 1.

Cole, K.S. and J.W. Moore. (1960). Ionic current measurements in the squid giant axon membrane. J. Gen. Physiol. 44: 123.

Condrea, E., P. Rosenberg and W.D. Dettbarn. (1967). Demonstration of phospholipid splitting as the factor responsible for increased permeability and block of axonal conduction induced by snake venom. I. Study on lobster axons. Biochim. Biophys. Acta, 135: 669.

Conway, B.E. (1952). Electrochemical data. Amsterdam and New York: Elsevier.

Cooley, J., F. Dodge and H. Cohen. (1965). Digital computer solutions for excitable membrane models. J. Cell. Comp. Physiol. 66: 99.

Curtis, H.J. and K.S. Cole. (1937). Transverse electric impedance of Nitella. J. Gen. Physiol. 21: 189.

Curtis, H.J. and K.S. Cole. (1938). Transverse electric impedance of the squid giant axon. J. Gen. Physiol. 21: 757.

Curtis, H.J. and K.S. Cole. (1942). Membrane resting and action potentials from the squid giant axon. J. Cell. Comp. Physiol. 19: 135.

Davydov, B. (1935). Uber die Geschwindigkeitsverteilung der sich im elektrischen Felde Bewegenden Elektronen. Phys. Zeits. Sovjetunion, 8: 59.

Delcroix, J.L. (1960). Introduction to the theory of ionized gases. New York: Interscience.

Ehrenstein, G. and D.L. Gilbert. (1966). Slow changes of potassium permeability in the squid giant axon. Biophys. J. 6: 553.

Fishman, H.M. (1970). Leakage current in the squid axon membrane after application of TTX and TEA. Biophys. J. 10: 109a.

Frankenhaeuser, B. (1963). A quantitative description of potassium currents in myelinated nerve fibres of Xenopus laevis. J. Physiol. 169: 424.

Frankenhaeuser, B. and A.L. Hodgkin. (1957). The action of calcium on the electrical properties of squid axons. J. Physiol. 137: 217.

Fricke, H. (1925). The electrical capacity of suspensions with special reference to blood. J. Gen. Physiol. 9: 137.

Frumento, A.S. (1965). Sodium pump: Its electrical effects in skeletal muscle. Science, 147: 1442.

Green, H.S. (1952). The molecular theory of fluids. Amsterdam: North-Holland Publishing Company.

Gainer, H. (1967). Plasma membrane structure: Effects of hydrolases on muscle resting potentials. Biochim. Biophys. Acta, 135: 560.

George, E.P. and E.A. Johnson. (1961). Solutions of the Hodgkin-Huxley equations for squid axon treated with tetraethylammonium and in potassium-rich media. Aust. J. exp. Biol. 39: 275.

Gilbert, D.L. and G. Ehrenstein. (1965). Effect of calcium and magnesium on voltage clamped squid axons immersed in isosmotic potassium chloride. Abstracts, 23rd Int. Cong. Physiol. Sci., #148.

Gilbert, D.L. and G. Ehrenstein. (1969). Effect of divalent cations on potassium conductance of squid axons: Determination of surface charge. Biophys. J. 9: 447.

Glynn, I.M. (1964). The action of cardiac glycosides on ion movements. Pharm. Rev. 16: 381.

Goldman, D.E. (1943). Potential, impedance and rectification in membranes. J. Gen. Physiol. 27: 37.

Gorter, E. and F. Grendel. (1925). On bimolecular layers of lipoids on the chromocytes of the blood. J. Exp. Med. 41: 439.

Hendler, R.W. (1971). Biological membrane ultrastructure. Physiol. Rev. 51: 66.

Hille, B. (1966). Common mode of action of three agents that decrease the transient change in sodium permeability in nerves. Nature, 210: 1220.

Hille, B. (1967). The selective inhibition of delayed potassium currents in nerve by tetraethylammonium ion. J. Gen. Physiol. 50: 1287.

Hille, B. (1968). Charges and potentials at the nerve surface. J. Gen. Physiol. 51: 221.

Hille, B. (1970). Ionic channels in nerve membranes. Progr. Biophys. Mol. Biol. 21: 1.

Hirschfelder, J.O., C.F. Curtiss and R.B. Bird. (1964). Molecular theory of gases and liquids. New York: Wiley and Sons.

Hodgkin, A.L. (1939). The relation between conduction velocity and the electrical resistance outside a nerve fibre. J. Physiol. 94: 560.

Hodgkin, A.L. (1951). The ionic basis of electrical activity in nerve and muscle. Biol. Rev. 26: 339.

Hodgkin, A.L. (1958). Ionic movements and electrical activity in giant nerve fibres. Proc. Roy. Soc. Lond. 148B: 1.

Hodgkin, A.L. (1964). The conduction of the nervous impulse. Springfield, Ill.: C.C. Thomas Pub. Co.

Hodgkin, A.L. and P. Horowicz. (1959). Movements of Na and K in single muscle fibres. J. Physiol. 145: 405.

Hodgkin, A.L. and A.F. Huxley. (1952a). Currents carried by sodium and potassium ions through the membrane of the giant axon of *Loligo*. *J. Physiol.* *116*: 449.

Hodgkin, A.L. and A.F. Huxley. (1952b). The components of membrane conductance in the giant axon of *Loligo*. *J. Physiol.* *116*: 473.

Hodgkin, A.L. and A.F. Huxley. (1952c). The dual effect of membrane potential on sodium conductance in the giant axon of *Loligo*. *J. Physiol.* *116*: 497.

Hodgkin, A.L. and A.F. Huxley. (1952d). A quantitative description of membrane current and its application to conduction and excitation in nerve. *J. Physiol.* *117*: 500.

Hodgkin, A.L., A.F. Huxley and B. Katz. (1949). Ionic currents underlying activity in the giant axon of the squid. *Arch. Sci. Physiol.* *III*: 129.

Hodgkin, A.L., A.F. Huxley and B. Katz. (1952). Measurement of current-voltage relations in the membrane of the giant axon of *Loligo*. *J. Physiol.* *116*: 424.

Hodgkin, A.L. and B. Katz. (1949). The effect of sodium ions on the electrical activity of the giant axon of the squid. *J. Physiol.* *108*: 37.

Hodgkin, A.L. and R.D. Keynes. (1955a). Active transport of cations in giant axons from *Sepia* and *Loligo*. *J. Physiol.* *128*: 28.

Hodgkin, A.L. and R.D. Keynes. (1955b). The potassium permeability of a giant nerve fibre. *J. Physiol.* *128*: 61.

Hodgkin, A.L. and W.A.H. Rushton. (1946). The electrical constants of a crustacean nerve fibre. *Proc. Roy. Soc. Lond.* *133B*: 444.

Horowicz, P. and C.J. Gerber. (1965a). Effects of external potassium and strophanthidin on sodium fluxes in frog striated muscle. *J. Gen. Physiol.* *48*: 489.

Horowicz, P. and C.J. Gerber. (1965b). Effects of sodium azide on sodium fluxes in frog striated muscle. *J. Gen. Physiol.* *48*: 515.

Hoyt, R.C. (1963). The squid giant axon. *Biophys. J.* *3*: 399.

Hoyt, R.C. (1968). Sodium inactivation in nerve fibers. *Biophys. J.* *8*: 1074.

Hoyt, R.C. and W.J. Adelman. (1970). Sodium inactivation. Experimental test of two models. *Biophys. J.* *10*: 610.

Huxley, A.F. (1959). Ion movements during nerve activity. *Ann. N.Y. Acad. Sci.* *81*: 221.

Johnston, T.W. (1960). Cartesian tensor scalor product and spherical harmonic expansions in Boltzmann's equation. *Phys. Rev.* *120*: 1103.

Johnston, T.W. (1966). General spherical harmonic tensors in the Boltzmann equation. *J. Math. Phys.* (Cambridge, Mass.) *7*: 1453.

Kao, C.Y. (1966). Tetrodotoxin, saxitoxin and their significance on the study of excitation phenomena. *Pharm. Rev.* *18*: 997.

Ketelaar, J.A.A. (1953). Chemical constitution: An introduction to the theory of the chemical bond. Amsterdam and New York: Elsevier.

Keynes, R.D. and R.C. Swan. (1959). The effect of external sodium concentration on the sodium fluxes in frog skeletal muscle. J. Physiol. 147: 591.

Langmuir, I. (1917). The constitution and fundamental properties of solids and liquids. II. Liquids. J. Am. Chem. Soc. 39: 1848.

Latimer, W.M. (1952). The oxidation states of the elements and their potentials in aqueous solutions. 2n ed. New Jersey: Prentice Hall.

Lecar, H., G. Ehrenstein, L. Binstock and R.E. Taylor. (1967). Removal of potassium negative resistance in perfused squid giant axons. J. Gen. Physiol. 50: 1499.

Ling, G.N. (1962). A physical theory of the living state. New York: Blaisdell Pub. Co.

Ling, G. and R.W. Gerard. (1949). The normal membrane potential of frog sartorius fibers. J. Cell. Comp. Physiol. 34: 383.

Lorentz, H.A. (1952). Theory of electrons and its applications to the phenomena of light and radiant heat. New York: Dover.

Mackey, M.C. (1968). Excitable membrane models: Statistical mechanical analysis of current-electric field relationships. Dissertation, University of Washington, Seattle.

Mackey, M.C. (1971a). Kinetic theory model for ion movement through biological membranes. I. Field-dependent conductances in the presence of solution symmetry. Biophys. J. 11: 75.

Mackey, M.C. (1971b). Kinetic theory model for ion movement through biological membranes. II. Interionic selectivity. Biophys. J. 11: 91.

Mackey, M.C. and M.L. McNeel. (1971c). Kinetic theory model for ion movement through biological membranes. III. Steady-state electrical properties with solution asymmetry. Biophys. J. 11: 664.

Mackey, M.C. and M.L. McNeel. (1971d). The independence principle. A reconsideration. Biophys. J. 11: 675.

Mackey, M.C. and M.L. McNeel. (1973). Determinants of time-dependent membrane conductance: The non role of classical ion-membrane molecule interactions. Biophys. J. 13: 733.

Marmont, G. (1949). Studies on the axon membrane. I. A new method. J. Cell. Comp. Physiol. 34: 351.

Mauro, A., F. Conti, F. Dodge and B. Schor. (1970). Subthreshold behavior and phenomenological impedance of the squid giant axon. J. Gen. Physiol. 55: 497.

Meves, H. and W.K. Chandler. (1965). Ionic selectivity in perfused giant axons. J. Gen. Physiol. 48: 31.

Moelwyn-Hughes, E.A. (1949). Ionic hydration. Proc. Cambridge Phil. Soc. 45: 477.

Moore, J.W. (1959). Excitation of the squid axon membrane in isosmotic potassium chloride. Nature, 183: 265.

Moore, J.W. and K.S. Cole. (1960). Resting and action potentials of the squid giant axon in vivo. J. Gen. Physiol. 43: 961.

Moore, J.W. and K.S. Cole. (1963). Voltage clamp techniques in Physical techniques in biological research, VI: 263.

Moore, J.W., N. Anderson, M. Blaustein, M. Takata, J. Lettvin, W.F. Pickard, T. Bernstein and J. Pooler. (1966). Alkali cation selectivity of squid axon membrane. Ann. N.Y. Acad. Sci. 137: 818.

Moore, J.W., M. Blaustein, N. Anderson and T. Narahashi. (1967). Basis of tetrodotixin's selectivity in blockage of squid axons. J. Gen. Physiol. 50: 1401.

Moore, J.W., T. Narahashi and T.I. Shaw. (1967). An upper limit to the number of sodium channels in nerve membrane? J. Physiol. 188: 99.

Moore, J.W., T. Narahashi and W. Ulbricht. (1964). Sodium conductance shift in an axon internally perfused with a sucrose and low-potassium solution. J. Physiol. 172: 163.

Morrone, T. (1968). Spherical-harmonic and power-series expansion of the Boltzmann equation. Phys. Fluids, 11: 1227.

Morse, P.M., W.P. Allis and E.S. Lemar. (1935). Velocity distributions for elastically colliding electrons. Phys. Rev. 48: 412.

Mueller, P., D.O. Rudin, H.T. Tien and W.C. Wescott. (1962). Reconstitution of excitable cell membrane structure in vitro. Circulation, XXVI: 1167.

Mulliken, R.S. (1933). Electronic structure of polyatomic molecules and valence period. V. Molecules RX_n. J. Chem. Phys. 1: 492.

Mullins, L.J. (1968). Ion fluxes in dialyzed squid axons. J. Gen. Physiol. 51: 146.

Mullins, L.J. and M.Z. Awad. (1965). The control of the membrane potential of muscle fibers by the sodium pump. J. Gen. Physiol. 48: 761.

Mullins, L.J. and F.J. Brinley. (1967). Some factors influencing sodium extrusion by internally dialyzed squid axons. J. Gen. Physiol. 50: 2333.

Mullins, L.J. and A.S. Frumento. (1963). The concentration dependence of sodium efflux from muscle. J. Gen. Physiol. 46: 629.

Mullins, L.J. and K. Noda. (1963). The influence of sodium-free solutions on the membrane potential of frog muscle fibers. J. Gen. Physiol. 47: 117.

Nakamura, Y., S. Nakajima and H. Grundfest. (1965). The action of tetrodotoxin on electrogenic components of squid giant axons. J. Gen. Physiol. 48: 985.

Narahashi, T. (1963). Dependence of resting and action potentials on internal potassium in perfused squid giant axons. J. Physiol. 169: 91.

Narahashi, T., N.C. Anderson and J.W. Moore. (1967). Comparison of tetrodotoxin and procaine in internally perfused squid giant axons. J. Gen. Physiol. 50: 1413.

Narahashi, T., T. Deguchi, N. Urakawa and Y. Ohkubo. (1960). Stabilization and rectification of muscle fiber membrane by tetrodotoxin. Am. J. Physiol. 198: 934.

Narahashi, T., J.W. Moore and R.N. Poston. (1967). Tetrodotoxin derivatives: Chemical structure and blockage of nerve membranr conductance. Science, 156: 976.

Narahashi, T., J.W. Moore and W.R. Scott. (1964). Tetrodotoxin blockage of sodium conductance increase in lobster giant axons. J. Gen. Physiol. 47: 965.

Narahashi, T. and J.M. Tobias. (1964). Properties of axon membrane as affected by cobra venom, digitonin and proteases. Am. J. Physiol. 207: 1441.

Nernst, W. (1888). Zur Kinetik der Lösung befindlichen Körper: Theorie der Diffusion. Z. Physik. Chem. 2: 613.

Nernst, W. (1889). Die elekromotorische Wirksamkeit der Ionen. Z. Physik. Chem. 4: 129.

Noble, D. (1966). Applications of Hodgkin-Huxley equations to excitable tissues. Physiol. Rev. 46: 1.

Overton, E. (1899). Ueber die allgemeinen osmotischen Eigenschaften der Zelle, ihre vermutlichen Ursachen und ihre Bedeutung für die Physiologie. Vjochr. Naturf. Ges. Zurich, 44: 88.

Pickard, W.F. (1969). Estimating the velocity of propogation along myelinated and unmyelinated fibers. Math. Biosciences, 5: 305.

Planck, M. (1890). Uber die Erregung von Elektricität und Wärme in Elektrolyten. Ann. Phys. Chem. Nevefolge, 39: 161.

Planck, M. (1890). Uber die Potentialdifferenz zwischen zwei verdünnten Lösungen binärer Elektrolyte. Ann. Phys. Chem. Nevefolge, 40: 561.

Rall, W. (1959). Branching dendritic trees and motoneuron membrane resistivity. Exptl. Neurol. 1: 491.

Rall, W. (1960). Membrane potential transients and membrane time constant of motoneurons. Exptl. Neurol. 2: 503.

Rall, W. (1962). Electrophysiology of a dendritic meuron model. Biophys. J. 2: 145.

Rall, W. (1962). Theory of physiological properties of dendrites. Ann. N.Y. Acad. Sci. 96: 1071.

Rall W. (1964). Theoretical significance of dendritic trees for neuronal input-output relations, In Neural theory and modeling, 73. Stanford University Press.

Rall, W. (1967). Distinguishing theoretical synaptic potentials, computed for different soma-dendritic distributions of synaptic output. J. Neurophysiol. 30: 1138.

Rice, S.A. and P. Gray. (1965). The statistical mechanics of simple liquids. New York: Interscience.

Roberts, G.E. and H. Kaufman. (1966). Tables of Laplace transforms. Philadelphia: W.B. Saunders.

Robertson, J.D. (1960). The molecular structure and contact relationships of cell membranes. Prog. Biophys. 10: 343.

Rojas, E., I. Atwater and F. Bezanilla. (1970). Fixed charge structure of the squid axon membrane, In Permeability and function of biological membranes, 273. Amsterdam: North-Holland Publishing Co.

Rojas, E. and G. Ehrenstein. (1965). Voltage clamp experiments on axons with potassium as the only internal and external cation. J. Cell. Comp. Physiol. 66: 71.(Suppl. 2).

Rojas, E., R.E. Taylor, I. Atwater and F. Bezanilla. (1969). Analysis of the effects of calcium or magnesium on voltage-clamp currents in perfused squid axons bathed in solutions of high potassium. J. Gen. Physiol. 54: 532.

Rushton, W.A.H. (1951). A theory of the effects of fibre size in medullated nerve. J. Physiol. 115: 101.

Sandbloom, J. (1972). Anomalous reactances in electrodiffusion systems. Biophys. J. 12: 1118.

Senft, J.P. (1967). Effects of some inhibitors on the temperature-dependent component of resting potential in lobster axon. J. Gen. Physiol. 50: 1835.

Senft, J.P. and W.J. Adelman. (1967). Ionic conductances in intact and internally perfused squid axons. Abstract WCZ, Biophysical Society, 11th Annual Meeting.

Sjodin, R.A. and L.A. Beauge. (1967). The ion selectivity and concentration dependence of cation coupled active sodium transport in squid giant axons. Current topics in Modern Biology 1:105.

Sjodin, R.A. and L.A. Beauge. (1968). Coupling and selectivity of sodium and potassium transport in squid giant axons. J. Gen. Physiol. 51: 152.

Skou, J.C. (1965). Enzymatic basis for active transport of Na^+ and K^+ across cell membranes. Physiol. Rev. 45: 596.

Stampfli, R. (1954). Saltatory conduction in nerve. Physiol. Rev. 34: 101.

Stillman, H., D.L. Gilbert and R.J. Lipicky. (1971). Effect of external pH upon the voltage dependent currents of the squid giant axon. Abstract TPM-F14, Biophysical Society, 15th Annual Meeting.

Takata, M., J.W. Moore, C.Y. Kao and F.A. Fuhrman. (1966). Blockage of sodium conductance increase in lobster giant axon by tarichatoxin (tetrodotoxin). J. Gen. Physiol. 49: 977.

Takenaka, T. and S. Yamagishi. (1966). Intracellular perfusion of squid giant axons with protease solution. Proc. Jap. Acad. 42: 521.

Tasaki, I. (1953). Nervous transmission. Springfield, Ill.: C.C. Thomas Pub. Co.

Tasaki, I. (1959). Demonstration of two stable states of the nerve membrane in potassium rich media. J. Physiol. 148: 306.

Tasaki, I. and S. Hagiwara. (1957). Demonstration of two stable potential states in the squid giant axon under tetraethylammonium chloride. J. Gen. Physiol. 40: 859.

Tasaki, I. and M. Shimamura. (1962). Further observations on resting and action potential of intracellularly perfused squid axon. Proc. Nat. Acad. Sci. 48: 1571.

Tasaki, I., I. Singer and A. Watanabe. (1965). Excitation of internally perfused squid giant axons in sodium-free media. Proc. Nat. Acad. Sci. 54: 763.

Tasaki, I., I. Singer and A. Watanabe. (1966). Excitation of squid giant axons in sodium-free external media. Am. J. Physiol. 211: 746.

Taylor, R.E. (1963). Cable theory, In Techniques in biological research, VI: 219. New York: Academic Press.

Taylor, R.E., Moore, J.W. and K.S. Cole. (1960). Analysis of certain errors in squid axon voltage clamp measurements. Biophys. J. 1: 161.

Thomas, R.C. (1972). Electrogenic sodium pump in nerve and muscle cells. Physiol. Rev. 52: 563.

Ussing, H. (1949). Ion transport across biological membranes. Acta Physiol. Scand. 19: 43.

Verwey, E.J.W. and J.T. Overbeck. (1948). Theory of the stability of lyophobic colloids. New York: Elsevier.

Villegas, R., F.V. Barnola, G. Camejo and H. Davila. (1970). Similarities between the requirements for the interaction of tetrodotoxin with the axon membrane and with cholesterol monolayers. Biophys. J. 10: 113a.

Watanabe, A., I. Tasaki, I. Singer and L. Lerman. (1967). Effects of tetrodotoxin on excitability of squid giant axons in sodium-free media. Science, 155: 95.

Weidman, S. (1950). Die Natur des induktiven Elements in biologischen Membranen. Experientia, VI: 1.

Woodbury, J.W. (1965). In Physiology and biophysics, ed. by T.C. Ruch and H.O. Patton, pp. 1 58. Philadelphia: Saunders.

Woodbury, J.W., S.H. White, M.C. Mackey, W.L. Hardy and D.B. Chang. (1970). Bioelectrochemistry, In Electrochemistry, 9B: 903. New York: Academic Press.

Woodward, R.B. (1964). The structure of tetrodotoxin. Pure Appl. Chem. 9: 49.

Appendix 1. The dimensions, derived MKS units, and defining equations for fundamental quantities used in the text.

Quantity	Dimensions	Derived MKS Unit	Defining Equation
Mass	M	Kilogram	
Length	L	meter	
Time	T	second	
Force	MLT^{-2}	newton	$F = ma$
Energy	ML^2T^{-2}	joule	$u = \int \underset{\sim}{F}.d\underset{\sim}{s}$
Charge	Q	coulomb	
Current	QT^{-1}	ampere	$j = dq/dt$
Volume charge density	QL^{-3}	coulomb/meter^3	$p = \lim_{\Delta V \to 0} \Delta q/\Delta V$
Surface charge density	QL^{-2}	coulomb/meter^2	$\sigma = \lim_{\Delta A \to 0} \Delta q/\Delta A$
Current density	$QT^{-1}L^{-2}$	ampere/meter^2	$J = \lim_{\Delta A \to 0} \Delta j/\Delta A$
Electric field	$MLQ^{-1}T^{-2}$	newton/coulomb	$\underset{\sim}{E} = \lim_{\Delta q \to o} \Delta \underset{\sim}{F}/\Delta q$
Potential	$ML^2Q^{-1}T^{-2}$	volt	$V = \int \underset{\sim}{E}.d\underset{\sim}{s}$
Capacitance	$Q^2T^2M^{-1}L^{-2}$	farad	$C = q/V$
Dipole moment	QL	coulomb-meter	$\underset{\sim}{\mu} = q\delta\underset{\sim}{l}$
Permittivity	$Q^2T^2M^{-1}L^{-3}$	coulomb^2/newton-meter^2	$\varepsilon_0 = 10^{-9}/36\pi$
Resistance	$ML^2Q^{-2}T^{-1}$	ohm	$R = V/j$
Conductance	$Q^2TM^{-1}L^{-2}$	mho	$G = j/V$
Resistivity	$ML^3Q^{-2}T^{-1}$	ohm-meter	$\rho = E/J$
Conductivity	$Q^2TM^{-1}L^{-3}$	mho/meter	$\sigma = J/E$

Appendix 2. Some useful physical constants in the MKS International Unit System.

Constant	Symbol	Value	Units
Speed of light in a vacuum	c	2.9979×10^{8}	meter/second
Elementary charge	e	1.6021×10^{-19}	coulomb
Avogadros constant	N	6.0225×10^{23}	$mole^{-1}$
Proton rest mass	m_p	1.6725×10^{-27}	Kilogram
Faradays constant	F	9.6487×10^{4}	coulomb/mole
Gas constant	R	8.3143	joule/^{o}Kelvin-mole
Boltzmanns constant	K	1.3805×10^{-23}	joule/^{o}Kelvin

Editors: K. Krickeberg; R.C. Lewontin; J. Neyman; M. Schreiber

Biomathematics

Vol. 1: **Mathematical Topics in Population Genetics**
Edited by K. Kojima
55 figures. IX, 400 pages. 1970
ISBN 3-540-05054-X

This book is unique in bringing together in one volume many, if not most, of the mathematical theories of population genetics presented in the past which are still valid and some of the current mathematical investigations.

Vol. 2: E. Batschelet
Introduction to Mathematics for Life Scientists
200 figures. XIV, 495 pages. 1971
ISBN 3-540-05522-3

This book introduces the student of biology and medicine to such topics as sets, real and complex numbers, elementary functions, differential and integral calculus, differential equations, probability, matrices and vectors.

M. Iosifescu; P. Tautu
Stochastic Processes and Applications in Biology and Medicine

Vol. 3: Part 1: Theory
331 pages. 1973
ISBN 3-540-06270-X

Vol. 4: Part 2: Models
337 pages. 1973
ISBN 3-540-06271-8

Distribution Rights for the Socialist Countries: Romlibri, Bucharest

This two-volume treatise is intended as an introduction for mathematicians and biologists with a mathematical background to the study of stochastic processes and their applications in medicine and biology. It is both a textbook and a survey of the most recent developments in this field.

Vol. 5: A. Jacquard
The Genetic Structure of Populations
Translated by B. Charlesworth; D. Charlesworth
92 figures. Approx. 580 pages. 1974
ISBN 3-540-06329-3

Population genetics involves the application of genetic information to the problems of evolution. Since genetics models based on probability theory are not too remote from reality, the results of such modeling are relatively reliable and can make important contributions to research. This textbook was first published in French; the English edition has been revised with respect to its scientific content and instructional method.

Springer-Verlag
Berlin
Heidelberg
New York

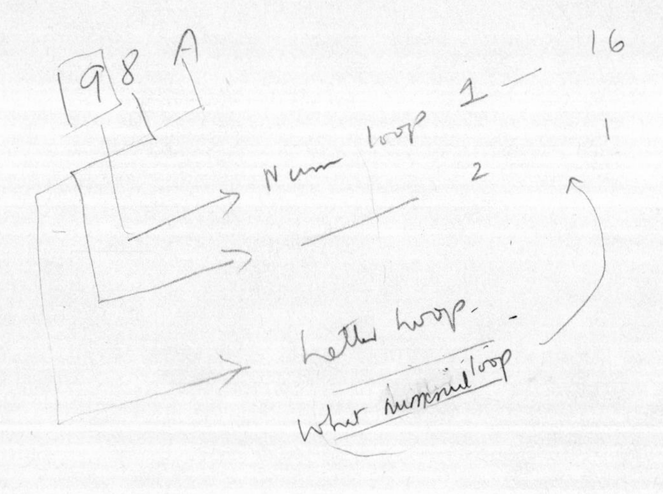
198A
16
1
Num loop
2
1
Letter loop.
Whole numeric loop